The study of ionic equilibria

The study of ionic equilibria

An introduction

Hazel Rossotti
Fellow and Tutor, St Anne's College, Oxford

Longman
London and New York

Longman Group Limited London

Associated companies, branches and representatives throughout the world

Published in the United States of America by Longman Inc., New York

© Longman Group Limited 1978

First published 1978

Library of Congress Cataloging in Publication Data

Rossotti, Hazel.
  The study of ionic equilibria.

  Bibliography: p.
  Includes index.
  1. Chemical equilibrium. 2. Ionic solutions.
I. Title.
QD503.R67   541.392   77-26048
ISBN 0-582-44175-7

Printed in Great Britain by
Richard Clay (The Chaucer Press) Ltd, Bungay, Suffolk

in memoriam Lars Gunnar Sillén

Contents

Acknowledgements

We are grateful to the following Publishers and respective authors for permission to reproduce copyright material:
Academic Press Inc. for a fig. from *Reference Electrodes* by G. J. Hills and D. J. G. Ives; Acta Chemica Scandinavia for Fig. 7 by F. T. C. Rossotti and H. S. Rossotti from Vol. 9 (1955), Fig. 1 by N. Ingri, G. Lagerstroni, M. Frydman and L. G. Sillén from Vol. 11 (1957), Figs. 1 and 2 by A. Olin from Vol. 14 (1960), Fig. 1 by Y. Sasaki from Vol. 15 (1961), Fig. 1 by K. A. Burkov, L. S. Lilic and L. G. Sillén and Fig. 1 by T. Sekini from Vol. 19 (1965), Figs. 1 and 2 by B. Norén from Vol. 21 (1967), Figs. 1 and 2 by I. Grenthe and G. Gardhammer from Vol. 23 (1969), Fig. 6 by D. Alexandersson and N. G. Vannerberg from Vol. 26 (1972), and Fig. 4 by R. Lundquist and J. Rydberg from Vol. A28 (1974) from *The Journal of Acta Chemica Scandinavia*; Addison-Wesley Publishing Company for Figs. 5.4, 5.6, 7.4, 7.7, 7.8, 7.11, 9.11, and 10.9 from *Ionic Equilibrium* by J. N. Butler; American Chemical Society for Fig. 4 by C. Tanford, S. A. Swanson and W. S. Shore from Vol. 77 (1955), Fig. 4 by Y. Nozaki, L. Bunville and C. Tanford from Vol. 81 (1959), Fig. 4 by T. Spiro and D. N. Hume from Vol. 83 (1961), Fig. 2 by D. L. Rabenstein from Vol. 95 (1973), and Fig. 1 by K. P. Anderson, E. A. Butler, D. H. Anderson and E. M. Woolley from Vol. 71 (1967) from *Journal of the American Chemical Society*; McGraw-Hill Book Company for Figs. 5.2 and 13.4 from *The Determination of Stability Constants* by F. J. C. Rossotti and H. S. Rossotti. © McGraw-Hill Book Co. (1961). Used with permission of McGraw-Hill Book Co.; Pergamon Press Ltd. for Fig. 8.8 by M. T. Beck from *Journal Inorg. Nucl. Chem*; Plenum Publishing Corp. for Figs. 9.5 and 10.6 by W. B. Guenther from *Chemical Equilibrium* (1975); Royal Swedish Academy of Science for Fig. 1 by Y. Sasaki and L. G. Sillén from *Arkiv for Kemi* Vol. 19 (1968); Van Nostrand Reinhold Company Ltd. for Figs. 9.1 and 10.2 by H. S. Rossotti from *Chemical Applications of Potentiometry* (1969); John Wiley & Sons Inc. for Figs. 3.5 and 11.5 by C. F. Baes and R. E. Mesmer from *The Hydrolysis of Cations* (1976), and Fig. 29.1(d) by C. Tanford from *The Physical Chemistry of Macromolecules* (1961).
We have been unable to trace the copyright owners of Fig. 2 by H. M. N. H. Irving from *Solvent Extraction Chemistry* by D. Dyrssen, J. O. Liljenzin and J. Rydberg, and any information which would enable us to do so would be appreciated.

Preface

I hope that this book may be of use to those who wish, or need, to acquire some knowledge of the methods used to study chemical equilibria in solution. There are already available a number of single chapters, or single sections, in textbooks of general physical chemistry, of electrochemistry or of metal complex chemistry. Such treatments are necessarily very brief and are usually restricted either to (often monobasic) Brönsted acids or to binary mononuclear metal complexes. Readers who seek a fuller account of the subject have little at their disposal other than very substantial review articles or books in the 'learned monograph' class, and they could be forgiven for finding these both unpalatable and indigestible. I have tried to promote the flow of understanding between these two widely separated levels by inserting a small, intermediate 'impurity band'.

It is my firm conviction that the algebraic description of stepwise equilibria is much easier than it looks. Since the underlying ideas, once assimilated, can readily be applied to a wide variety of chemical equilibria, I have tried to show how very closely the methods for studying ternary polynuclear metal complexes, or the binding of cations by proteins, are related to procedures for investigating very simple equilibria, such as the dissociation of a monobasic acid. The mathematics should cause no trouble to anyone who has reached G.C.E. O-level 'additional' standard, or its equivalent outside the UK. References to original papers are not included; any reader who wishes to go orienteering in the vast literature on equilibrium chemistry can start from a learned monograph, of which several are included in the list of suggestions for further reading, at the end of the text. For those with an urge to try their hand at equilibrium constant correlations there are rich deposits in the data compilations, which are also listed.

Although the book was written primarily for students following an honours course in chemistry, I hope that other groups of reader may also find it useful. For example, the earlier chapters on both acid–base equilibria and metal complexes could appropriately be incorporated into less specialised courses. But the book need not be restricted to formal work and may well form useful background reading for physical and inorganic chemists, biochemists and analysts who are preparing to work on solution equilibria. The chapters outlining methods of calculating equilibrium constants and ways of representing equilibria were written mainly for the benefit of such readers, although I hope it may also form the basis of some post-graduate course-work.

I am, of course, very grateful to all those who helped the book on its way: to my undergraduate pupils who, albeit unwittingly, stimulated me to to try to refute the myth that equilibrium chemistry is difficult; to Mrs Inga-Britta Currie and Mrs Heather Holloway who transformed displeasing drafts into immaculate typescript; to Dr Alan Sharpe, for his kindly but critical eye; to copyright holders, for their generous permission to reproduce illustrations; and to the publishers, for a happily uneventful collaboration.

H.S.R.,
St Anne's College
Oxford.
September 1977.

List of main symbols

Symbols for species (e.g. an anion A) are written in Roman type. Symbols for analytical concentrations (e.g. total concentration A_t, initial concentration A_i, of A) are written in *italic* type. The equilibrium concentration of A is represented by [A] and the activity of A by {A}. Normalised variables are written in **bold** face.

A	anion
A_s	optical absorbency (equation 3.36)
B	base
C	cation
E	e.m.f.
$E^{\ominus}$	standard e.m.f.
$E^0, E^{0\prime}$	e.m.f. defined by equation (equation 3.14)
E_{cell}	observed e.m.f. of cell
E_p	half-cell potential of probe half-cell
E_{ref}	half-cell potential of reference half-cell
F	faraday
G	Gibbs free energy
H	hydrogen ion
[H]	$= \beta[H]$ (equation 3.33)
h	$= \beta_2^{\frac{1}{2}}[H]$ (equation 4.32)
i_S	intensive factor for species S
I	ion
I	ionic strength
j	number of protons bound to a base
$\bar{j}$	average value of j
J	maximal value of j
K	equilibrium constant
K_i	'intrinsic' constant for ion–polymer equilibria
K_j	stepwise formation constant of acid H_jA
K_M	selectivity coefficient of membrane
K_n	stepwise stability constant of complex ML_n
K_s	solubility product
K_T	tautomerisation constant
K_w	ionic product of water
K_1^0, K_1^{+-}	microscopic formation constants of acids HA^0 and HA^{+-}

K_2^0, K_2^{+-}	microscopic formation constants of acid H_2A^+ from HA^0 and HA^{+-}
l	optical path length
L	ligand
M	metal ion
M	(after number) molarity $=$ mol dm^{-3}
n	number of ligands bound to a metal ion
$\bar{n}$	average value of n
N	maximal value of n
P_c	partition coefficient of species ML_c
q	distribution ratio
Q, QH$_2$	quinhydrone
$_aQ$, $_cQ$, $_\gamma Q$	quotient of activities, concentrations or activity coefficients
Q_c	quantity proportional to α_c
$\mathbf{Q}_c$	normalised value of Q_c
R	gas constant
$R_{c,r}$	ratio $[ML_c]/[ML_r]$ (equation 13.4)
S	species
S	solubility; or sum of concentrations (equation 10.10)
t_S	Hittorf transport number of S
T	degrees Kelvin
u_S, u_n	ionic mobility of S, or of ML_n
U	uncharged species
v, V	volumes
w_S	washburn number of S ($= t_S/z_S$)
X	auxiliary species
z	charge
$\bar{z}$	average value of z
α_c	fraction of A in the form of H_cA or of M in the form of ML_c
β_n	overall formation constant of ML_n
β_j^H	overall formation constant of H_jA
γ_S	activity coefficient of S
ε_S	extinction coefficient of S
$\mathscr{E}$	observed extinction coefficient
μ_S	chemical potential of S
ρ	$= K_1/K_2$

Right-hand subscripts:

a, b	of acid, base
c	of species H_cA or ML_c
diss	dissociation
e	equivalence point
i	initial
j	of acid H_jA
j, J	junction

L   left-hand side
M (MA, MC)   membrane (permeable to anions $_A$, or cations $_C$)
n of complex ML_n
o   organic phase
R   right-hand side; resin phase
S   of species S
t   total
u   uncharged
w   water

Part I

Introduction

Chapter 1

Chemical equilibria

This book is about chemical equilibria. Now equilibrium, as is well known to those who are acquainted with zodiacal lore or elementary Latin, is the state in which equal forces act on either side of the Libra, or scales. A balance point is thereby achieved.

Many of us first meet the concept of equilibrium in a mechanical context. Spheres rest in hollows, on horizontal surfaces and on humps; and a cone can rest on its base, on its side or, improbably, on its apex. We are traditionally exhorted to distinguish between the three different types of equilibrium by contemplating the result of 'the body's being subjected to a small displacement'.

The concept of chemical equilibrium retains the idea of balance between two opposing forces, but the forces in question are no longer the gravitational pulls on the two pans of the scales. In a chemical system, equilibrium is reached when the tendency of one or more substances to change, or react together, to form others is exactly balanced by the opposing tendency of these products to change back into the initial reactants. It may take some time for this equilibrium to be established; but once it has been reached, the quantities of both reactants and products remain constant. Unless matter is added to, or removed from, a system (or some condition such as temperature is changed) a chemical equilibrium stays indefinitely at its balance point in the same way that the two arms of a pair of scales, once balanced, remain poised for as long as they are undisturbed. It is, indeed, this constancy over a period of time which forms the basis of the similarity between a pair of scales and a chemical reaction. Discussions about the relative concentrations of reactants and products in terms of, say, the adjustable position of the suspension point of a single-pan balance would be intellectual self-indulgence, liable to place the analogy under undue strain.

Chemical equilibria may be set up in a large number of different types of system. They may involve homogeneous reactions in liquids or gases; or heterogeneous reactions in systems consisting of a gas in contact with one or more solids or of a liquid in contact with a gas, a second liquid or a solid. Most of the chemical equilibria which will be discussed in this book are those which occur in a homogeneous, and usually predominantly aqueous, solution. But reactions which take place between these solutions and an immiscible liquid or sparingly soluble solid will also be considered. Heterogeneous equilibria between different phases of a pure substance (e.g. melting, boiling and sublimation) will be excluded, as will heterogeneous equilibria which occur in the absence of liquid (e.g. the thermal decomposition of calcium carbonate or the formation of gaseous ammonia from its elements).

Chemical equilibria in aqueous solution are usually first encountered in the guise of the dissociation of a weak acid. We learn that dilute acetic acid solution contains (hydrated) hydrogen ions, acetate ions and undissociated molecules of the acid. In a solution of acetic acid of concentration 10^{-2}M (i.e. 10^{-2} mol dm^{-3}), only about 4 per cent of the molecules are dissociated; and the solution comes to equilibrium at this same balance point whether the solution is prepared by the dilution of concentrated acetic acid, or by adding hydrogen ions (from hydrochloric acid) to acetate ions (from sodium acetate). Since the acidic properties of a solution are determined by the presence of the hydrogen ion, we learn that a substance in which only a small proportion of molecules is dissociated into hydrogen ions is called a weak acid. Hydrocyanic acid is considerably weaker than acetic acid; in a 10^{-2}M solution, fewer than 0.03 per cent of its molecules are dissociated. In both acids, however, a chemical equilibrium is set up as a result of the opposing tendencies of the molecules to dissociate and the ions which are formed from them to recombine. Even though dissociation is slight, both the reactants and the products are present in detectable amounts and the balance point persists indefinitely provided that the solutions are undisturbed.

Our early learning about weak acids was doubtless accompanied by a definition of a strong acid as a substance which was completely, or almost completely, dissociated into hydrogen ions in aqueous solution. Hydrochloric and nitric acids provided familiar examples, and indeed both acids are so strong that no undissociated molecules are detectable in 10^{-2}M solutions of either substance in water. It might seem inappropriate to discuss such solutions in terms of chemical equilibria; there can hardly be said to be a balance point between reactants and products when the reactants have totally disappeared. But the solution is nonetheless at equilibrium, in the sense that its composition does not vary with time. It is merely that, for a very strong acid, the balance point lies so far towards complete dissociation that no detectable quantity of un-ionised molecules remains. Although our elementary chemical education promoted awareness of the fact that some acids were di- or even tri-protic, and even that they yielded their successive protons with varying enthusiasm, such knowledge was mainly set in the framework of stoichiometry rather than of equilibrium chemistry. We learned that one mole of phosphoric acid could consume three of sodium hydroxide, but we were not taxed with the problem of disentangling the position of chemical equilibrium in a solution of a sodium hydrogen phosphate. However, although elementary acid–base theory is of necessity somewhat restricted in scope, its limitations are of narrowness rather than of error. No later unlearning is needed; merely extension.

Our second point of early contact with chemical equilibria involved metal ion complexes. These are conventionally discussed in terms of the formation (rather than the dissociation) of complexes by the combination of a metal ion with an anion, such as chloride or thiocyanate, or with a small molecule, such as ammonia, thereby causing dramatic changes in the solubility of a precipitate or in the colour of a solution. We learn that silver chloride dissolves in ammonia to form the complex ion $Ag(NH_3)_2^+$ and that the deep blue colour caused by the addition of ammonia to a solution of copper sulphate is due to the formation of the copper tetrammine ion $Cu(NH_3)_4^{2+}$. Chemists who had some early training in volumetric analysis may have used 'ferric alum' as an indicator in Volhard's titration; when excess thiocyanate solution is added to a solution of a silver salt, the end-point is detected by the sudden appearance of a red colour, variously ascribed in school books to the species $FeSCN^{2+}$, $Fe(CNS)_3$ or $Fe(CNS)_6^{3-}$.

The equilibrium chemistry of metal complexes has had but poor treatment from the writers of textbooks compared with the careful, if sometimes tedious, consideration which acid–base equilibria have received. This book attempts to produce a more (evenly) balanced view. The metal complexes discussed include both the relatively simple ones mentioned above and the much bulkier species formed by combination of metal ions with large organic molecules or anions.

The ill-treatment of metal complex equilibria in elementary textbooks has varied from excessive brevity and misleading over-simplification to total neglect or plain error. Many metal 'complexes', such as the silver and copper ammines mentioned above, are of course no more complex than the components from which they are formed. Replacement of the water molecules in the primary hydration sphere of $Ag.2H_2O^+$ by two ammonia molecules does not lead to an obvious increase in complexity. The question might seem to be one of subjective semantics. A more radical complaint against representation of metal complexes in elementary (and some less elementary) textbooks is the almost complete disregard of the fact that, even on solely statistical grounds, a supply of complexing agent is likely to be shared out amongst the metal ions which are present. So it is unlikely that the ion $Cu(NH_3)_4^{2+}$ will be formed in appreciable concentration in any 'cuprammine' solution unless the total concentration of ammonia is more than three times that of the total concentration of copper ions. It would have been so easy to describe the development of the deep blue colour solution in terms of 'successive replacement of water in the hydrated copper ion by ammonia molecules, until, in the presence of a large excess of ammonia, the ion $Cu(NH_3)_4^{2+}$ is completely formed'. The foundation would then have been laid for a more quantitative approach at a later stage and nothing would have needed to be unlearned. But traditional textbooks have usually favoured the whole-hog approach of assuming that the metal ion is fully complexed.

To what extent are these pleas for verbal rigour sheer pedantry? Are lower complexes likely to be present in appreciable concentrations? Let us consider the silver ammine complexes, which are unusual in that the ion $AgNH_3^+$ is rather unstable relative to Ag^+ and $Ag(NH_3)_2^+$, with the result that all three species may be present in the same solution (see Fig. 1.1). If equal volumes of, say, 0·1M solutions of silver nitrate and ammonia are mixed, the equilibrium solution will contain roughly 40 per cent of 'uncomplexed' silver ions, 40 per cent of the whole-hog complex $Ag(NH_3)_2^+$ and 20 per cent of the intermediate species $AgNH_3^+$. (It is a peculiarity of the silver ammonia system that the proportion of the intermediate complex can never exceed 20 per cent.) If the volume of the ammonia solution is twice that of the silver nitrate, about 92 per cent of the silver is in the form of the higher complex.

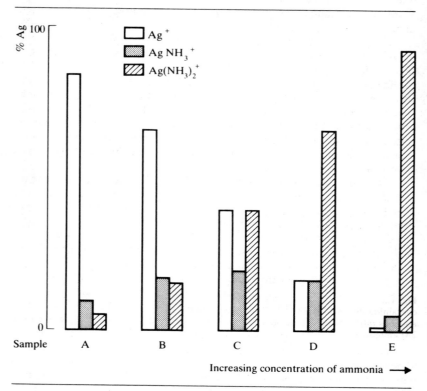

Fig. 1.1 Histograms showing the percentage of silver in various forms in solutions containing different amounts of ammonia. Samples are obtained by mixing 10 ml of aqueous 0·1M silver nitrate with x ml of aqueous 0·1M ammonia: A, $x = 2$; B, $x = 5$; C, $x = 10$; D, $x = 15$; E, $x = 20$.

A 0·1M excess of ammonia is needed before the proportion of the higher complex reaches 99·9 per cent. We should therefore attribute the solubility of silver chloride in aqueous ammonia to the formation of both the complexes $AgNH_3^+$ and $Ag(NH_3)_2^+$, their relative concentrations depending on the concentration of free ammonia in the solution.

In the same way, we may ask which complexes are present when ammonia has been added to a solution of copper sulphate. The approximate percentage of copper in each of the five most likely forms is shown in Fig. 1.2 for the dropwise addition of bench ammonia to a dilute solution of copper sulphate.

The highest complex is not completely formed, even in so dilute a copper solution as 10^{-2}M, until more than the same volume of 2M ammonia has

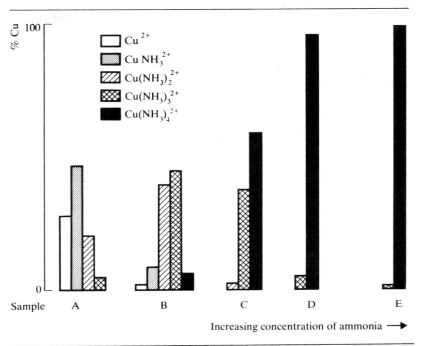

Fig. 1.2 Histograms showing the percentage of copper in various forms in solutions containing different amounts of ammonia. Samples are obtained by treating 10 ml of aqueous 0·01M copper sulphate solution with y drops (of volume 0·03 ml) of aqueous 2M ammonia: A, $y = 4$; B, $y = 4$; C, $y = 8$; D, $y = 25$; E, $y = 360$.

been added. Since all four ammine ions are a darker (and purpler) blue than the 'uncomplexed' copper ion, the addition of even one drop of bench ammonia does however cause a marked deepening of colour.

The third example of metal complex formation mentioned earlier involves the use of excess of the ion Fe^{3+} to detect the very small excess of thiocyanate ions present at the equivalence point of a titration. Common sense would suggest that, since the iron is present in large excess, the only iron thiocyanate complex which is present is $FeSCN^{2+}$; and this indeed is the case. The two other species mentioned together with complexes containing two, four and five thiocyanate groups per iron ion, are formed in solutions which contain much higher concentrations of the thiocyanate ion. But authors of early textbooks appear to have had a marked preference for the highest complex, or, failing that, the uncharged species.

In order to calculate the concentration of the various complexes which are present in a solution of a particular overall composition, or indeed even to ascertain which species are present, we need to know where the balance point of the appropriate chemical equilibrium lies. This book is about how we obtain such information, both for acids and for metal complexes. We shall first discuss briefly the origin of the balance point and describe experimental methods for studying very simple equilibria, such as the dissociation of acetic acid. We can then consider how some of these methods can be applied to more complicated equilibria, such as the dissociation of di- and poly-protic acids, the stepwise formation of metal complexes, and the binding of ions to proteins. We shall finally discuss mathematical methods for handling the experimental data and the purposes for which the results may be used.

Chapter 2

The balance point

Traditional balances are often designed so that, when equilibrium has been reached, a pointer sets in a vertical position, indicating that the force acting on one arm of the balance, multiplied by the distance between the pivot and that point at which the gravitational force acts on the arm, is exactly counterbalanced by analogous moment of the force acting on the other arm of the balance. Chemical equilibrium is similarly achieved when the driving force of a change in one direction is exactly opposed by the driving force of the same reaction in the reverse direction.

The driving force of a chemical process is, of course, the free energy change which accompanies it; a chemical system will tend to a state in which the free energy is minimal. So the driving force of a reaction is the difference between the free energy of the combined products and that of the original reactants. Suppose that we add a solution of dilute aqueous hydrochloric acid to an aqueous solution which contains sodium ions and anions A^-. If the sum of the free energies of the (hydrated) ions H^+ and A^- in the presence of the species Na^+, Cl^- and water is greater than the free energy of the undissociated acid HA (at the same concentration in the identical aqueous sodium chloride environment), some H^+ ions will combine with the anions to form molecules of HA. The free energy of the solution is thereby lowered. Formation of undissociated acid will proceed until there is no difference between the free energy of, on the one hand, the remaining H^+ and A^- ions, and, on the other, the molecules of HA which are generated. The exact position of chemical equilibrium is determined by the contributions which each of the participating species, H^+, A^- and HA, makes to the free energy. Not unnaturally, these contributions depend both on the nature, and on the concentration, of the species concerned as well as on temperature and pressure; and they may also be influenced, though to a lesser extent, by the presence of other species, such as the sodium ion.

The effect of one species, S, on the free energy of a chemical system may conveniently be discussed in terms of its chemical potential μ_S which is related to rate of change of the free energy of a system with the concentration of S at constant temperature, pressure and composition. (This last condition cannot, of course, be rigorously achieved unless all other species are present in infinitely large numbers; only then can we add ·S to a system without causing any appreciable change in the concentrations of the other species.)

The chemical potential μ_S of S in a particular solution is given by

$$\mu_S = \mu_S^{\ominus} + RT \ln \{S\} \tag{2.1}$$

where $\mu_s^{\ominus}$ is the chemical potential in some arbitrarily chosen standard state, R is the gas constant and T the absolute temperature. The activity {S} of S depends on the concentration [S] of S and to a lesser extent on the concentrations of all other species which are present. It is convenient to define the activity coefficient γ_s of S as the ratio

$$\gamma_s = \frac{\{S\}}{[S]} \qquad (2.2)$$

and to express the chemical potential as

$$\mu_s = \mu_s^{\ominus} + RT \ln [S] + RT \ln \gamma_s \qquad (2.3)$$

The three terms on the right-hand side of equation (2.3) represent the contribution to μ_s of the nature of S, the concentration of S, and the interaction of S ions or molecules with neighbouring particles.

A glance at equation (2.3) establishes that the value of {S} depends on the value of $\mu_s^{\ominus}$, which in turn is determined by the choice of standard state; and, in order that the activity coefficient γ_s be properly dimensionless, the values of {S} and S must be expressed on the same concentration scale. This book is concerned with equilibria in (at least moderately) dilute solution, at effectively constant pressure. Concentrations and activities under such conditions are frequently expressed as molarities, i.e. as moles of solute per cubic decimetre of solution. (Molarities, or moles of solute per kilogram solvent, are sometimes used in precise work. Concentrations expressed on the molal scale have the advantage of being independent of temperature.)

The standard state which is conventionally used for solutes may be envisaged by considering how the measurable quantity

$$\mu_s - RT \ln [S] = \mu_s^{\ominus} + RT \ln \gamma_s \qquad (2.4)$$

varies with [S]. As the concentration of S approaches zero, further decrease in [S] causes negligible change in the environment of the relatively few particles of S which are present in solution. The value of γ_s and hence that of ($\mu_s^{\ominus} + RT \ln \gamma_s$), then approach constancy. If the only species present are S and solvent, the value of γ_s tends to unity as S tends to zero. So, in very dilute solution,

$$\mu_s = \mu_s^{\ominus} + RT \ln [S] \qquad (2.5)$$

From equation (2.5) we see that $\mu_s = \mu_s^{\ominus}$ when [S] = 1, provided also that we can still take γ_s to be unity. The standard state of S is, therefore, a hypothetical solution, of unit concentration, which behaves as if the activity coefficient of S were unity. Since activity coefficients of ionic solutes in solutions of unit molarity (or of unit molality) differ markedly from unity, this standard state is truly hypothetical; indeed it is a mere mental construct. The standard state for the solvent is taken to be the pure liquid (which is, of course, at unit concentration on the mole fraction scale).

A solution which is infinitely dilute with respect to a particular solute S may nonetheless contain an appreciable concentration of other solute species. Equilibria are indeed often studied in aqueous solutions which contain a constant concentration of background electrolyte, such as $0 \cdot 1M$ KNO_3 or $3M$ $NaClO_4$. In 'constant ionic media' of this type, as in predominantly aqueous solutions, the quantity $\mu_S - RT \ln [S]$ becomes constant as [S] tends to zero. There are two obvious choices for the standard state in a constant ionic medium; and both are as hypothetical as the standard state in solutions without a background salt.

1. We may put $\gamma_S = 1$ when [S] = 0, and so define the standard state of S as a hypothetical solution of unit concentration of S in the particular ionic medium, with particles of S interacting with the medium exactly as they would if S were at infinite dilution (but by implication, not interacting with each other).
2. We may alternatively define the standard state of S as a hypothetical solution of unit concentration of S in pure solvent, interacting only with solvent molecules and in such a way as it would at infinite dilution. As the value of [S] tends to zero in the ionic medium, the value of γ_S again becomes constant. But it is not unity, as it reflects the difference between the behaviour of an infinitely dilute solution of S in pure solvent and that of an infinitely dilute solution of S in the ionic medium.

We may illustrate the use of chemical potentials by looking more closely at a solution containing a weak acid HA. The system will be in equilibrium when its free energy is minimal, and when the free energy change for the dissociation

$$HA \ aq \rightarrow H^+ \ aq + A^- \ aq \tag{2.6}$$

is zero. So at equilibrium

$$\mu_{HA} = \mu_{H^+} + \mu_{A^-} \tag{2.7}$$

(where the suffix 'aq' is omitted for simplicity). Invoking equation (2.1) we may write

$$(\mu_{H^+}^{\ominus} + \mu_{A^-}^{\ominus} - \mu_{HA}^{\ominus}) + RT \ln \frac{\{H^+\}\{A^-\}}{\{HA\}} = 0 \tag{2.8}$$

where the term in parentheses is the standard free energy change $\Delta G^{\ominus}$ which would accompany reaction (2.6) if all participants were in their standard states. The value of $\Delta G^{\ominus}$ is constant, since it depends only on the nature of the species involved. The activity quotient

$$_aQ = \frac{\{H^+\}\{A^-\}}{\{HA\}} \tag{2.9}$$

must therefore also be constant at equilibrium. Since

$$\{S\} = [S]\gamma_S \tag{2.2}$$

the concentration quotient

$$_cQ = \frac{[H^+][A^-]}{[HA]} \qquad (2.10)$$

will also be constant if, but only if, the activity coefficient quotient

$$_\gamma Q = \frac{\gamma_{H^+}\gamma_{A^-}}{\gamma_{HA}} \qquad (2.11)$$

can be taken as constant (e.g. at very low concentrations of H^+, A^- and HA in a salt medium) or even as unity (in effectively pure solvent). The quantities $_aQ$ and $_cQ$ are indeed known as equilibrium constants and are often denoted by K, or β, embellished with various sub- and super-scripts which define the quantities and reactions to which they refer (see sect. 3.1).

Although many of the equilibria which are discussed later are more complicated than the dissociation of a monobasic acid, they may be treated in exactly the same way. We may represent the general reaction

$$a A + b B \cdots \rightarrow \cdots y Y = z Z \qquad (2.12)$$

by the even more general formula

$$0 \rightarrow \sum_S \nu_S S \qquad (2.13)$$

where the stoichiometric coefficients ν_S are positive for products and negative for reactants. Thus ν_S represents $-a$, $-b$, $-y$ and z when S is respectively A, B, Y and Z. Since $\Delta G = 0$ at equilibrium, equations (2.2), (2.8), (2.9) and (2.10) can be combined to give

$$\Delta G^{\ominus} = \sum_S \nu_S \mu_S = RT \sum_S \nu_S \ln \{S\} \qquad (2.14)$$

$$= RT \ln {}_aQ \qquad (2.15)$$

$$= RT \ln {}_cQ_\gamma Q \qquad (2.16)$$

where the activity quotient

$$_aQ = \frac{\{Z\}^z\{Y\}^y \cdots}{\{A\}^a\{B\}^b \cdots} = \prod \{S\}^{\nu_S} \qquad (2.17)$$

is the equilibrium constant for reaction (2.12), and where the analogous concentration quotient $_cQ$ is also constant under conditions when $_\gamma Q$ is constant.

Clearly, the larger the value of the equilibrium constant, K ($= {}_aQ$ or $_cQ$), the farther to the right lies the balance point of the reaction. Thus for the reaction

$$A + B \rightarrow Y + Z \qquad (2.18)$$

1 per cent conversion of reactants to products gives $K = 0.000\,102$; 50 per cent conversion gives $K = 1$ and 99 per cent conversion gives $K = 9801$.

Since K changes by nearly 10^8 for even this modest difference in chemical behaviour it is usually convenient to express equilibrium constants on a logarithmic scale (see Fig. 2.1).

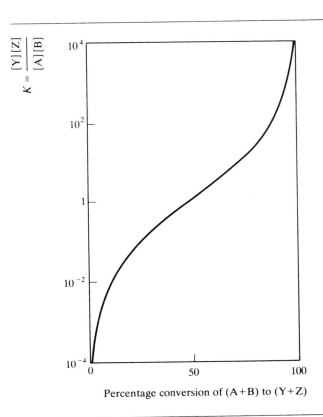

Percentage conversion of (A+B) to (Y+Z)

Fig. 2.1 The relationship between the value of the equilibrium constant and the position of equilibrium for the reaction $A + B \rightarrow Y + Z$ using equal initial concentrations of the reactants.

The expression for the equilibrium constant was not originally derived in terms of free energy or chemical potential; indeed it predates these concepts. Rather, it was expressed in terms of the rate at which changes occur and it was rightly surmised that, at equilibrium, the rate at which products are formed is exactly equal to the rate at which they revert to the reactants. Moreover, the rate at which many reactions proceed was found to be proportional to the quantity we should now write as $\prod \{S\}^{rs}$ reactants. So, the rate of a forward reaction

$$a\,A + b\,B \rightarrow \tag{2.12a}$$

is given by

$$\mathbf{R_f} = k_f\{A\}^a\{B\}^b \tag{2.19}$$

while that of the back reaction

$$\leftarrow zZ + yY \tag{2.12b}$$

is given by

$$\mathbf{R_b} = k_b\{Z\}^z\{Y\}^y \tag{2.20}$$

where the proportionality factors k_f and k_b are termed rate constants. At equilibrium $\mathbf{R_f} = \mathbf{R_b}$, and

$$k_f\{A\}^a\{B\}^b = k_b\{C\}^c\{D\}^d \tag{2.21}$$

so that the activity quotient

$$_aQ = \frac{k_f}{k_b} \tag{2.22}$$

is indeed shown to be constant, at least for those processes for which the rate laws are given by equations (2.19) and (2.20).

The rest of this book will deal with how, and why, values of the equilibrium constants $_aQ$ and $_cQ$ are measured.

Part II

Acids and bases

Chapter 3

One-step protonation

We shall start by considering the ways in which we can study the equilibria involving the addition of one proton H to a base A, such as the acetate ion or the ammonia molecule. (In the interests of typesetting, easy reading and generality we shall not normally show charges on ions.)

3.1 A plethora of constants

Even these simple, one-step reactions may be alternatively represented as the dissociation

$$HA \rightarrow H + A \tag{3.1}$$

or the formation

$$H + A \rightarrow HA \tag{3.2}$$

of the acid HA. The equilibrium constant of reaction (3.1) is the acid dissociation constant of HA, conventionally written as K_a, K_{diss} or just K. That for the converse reaction (3.2) is the formation constant, or stability constant, of HA, usually represented by β_1, K_1 or, unfortunately, also by plain K. We shall normally discuss formation reactions because the algebra which has been evolved to describe them is more general and less cumbersome; but as many readers will be more at home with dissociation constants of monobasic acids, we shall, in this chapter only, give equations in terms of both dissociation constants (K_{diss}) and formation constants (β). We shall unfortunately need to distinguish between three subspecies of each, depending on whether they are:

(i) activity quotients

$$_aK_{diss} = \frac{\{H\}\{A\}}{\{HA\}} = \frac{1}{_a\beta} \tag{3.3}$$

(ii) concentration quotients

$$_cK_{diss} = \frac{[H][A]}{[HA]} - \frac{1}{_c\beta} \tag{3.4}$$

or (iii) Brönsted constants

$$_B K_{diss} = \frac{\{H\}[A]}{[HA]} = \frac{1}{_B \beta}$$ (3.5)

which are a cross between the two. We shall drop the subscript $_c$ whenever we can do so without courting ambiguity. The unadorned symbols K and β therefore represent concentration quotients. Since the protonation of an unchanged base, such as ammonia, often occurs in alkaline solution, it has in the past been discussed in terms of the hydrolytic reaction

$$A + H_2O \rightarrow HA + OH$$ (3.6)

In very dilute solution, the water is effectively in its standard state, and so the concentration quotient for reaction (3.6) may be written as $[HA][OH]/[A]$. It is known as the (concentration) hydrolysis constant, or basic dissociation constant, of A, and is often denoted, confusingly, as K_b. Obviously

$$K_b = K_w \beta = \frac{K_w}{K_{diss}}$$ (3.7)

where $K_w = [H][OH]$ is the (concentration) ionic product of water. Now the standard state of water is, by convention, taken to be the pure liquid. For very dilute solutions, we may therefore set $\{H_2O\} = 1$, and hence $[H_2O] = 1$. So the ionic product, K_w, of water is merely its acid dissociation constant.

Equilibrium constants analogous to those in equation (3.7) may, of course, also be expressed in activities rather than concentrations.

There seems to be an extravagant variety of equilibrium constants for describing so simple a process as the addition of one proton to one molecule or ion of base. We shall confine ourselves to the twin series β and K_{diss} (including K_w) since values of the basic dissociation constant may be calculated from them.

Acid–base equilibria are best studied by physical methods which give information about the position of equilibrium but which do not disturb it. The experimental variable may be either the concentration or activity of one of the participating species (often H), or some property of the solution as a whole. Some methods combine both approaches. The value of K_{diss} for a monobasic acid may, for example, be obtained from potentiometric measurements of pH, from measurements of the conductivity of the solution, or from measurement of optical absorbency as a function of pH. We shall assume that we know the total analytical concentration of a solution from the way it was prepared; so, if we are discussing an acetate buffer which has been prepared by adding hydrochloric acid to sodium acetate, we shall assume that we know the concentrations of sodium and chloride ions and the total concentrations, A_t and H_t, of acetate and of dissociable hydrogen ions (see sect. 5.2 and 7.6).

We shall consider the three methods mentioned above together with some partition methods, such as solubility and liquid–liquid partition. These techniques form the foundation of much of equilibrium chemistry. We shall start with pH methods, since they are conceptually the simplest, by far the commonest and, also if properly used, the best.

3.2 pH methods

We shall assume here that the *concentration* of hydrogen ions can be measured potentiometrically and discuss later whether or not this assumption is justified. Now we know the values of A_t and H_t; and we also know which species contribute to these quantities, although we do not know in what proportion they do so. So we may write the mass balance expressions

$$A_t = [A] + [HA] \tag{3.8}$$

and

$$H_t + [OH] = [H] + [HA] \tag{3.9}$$

(The term [OH] arises because the protonic species on the right-hand side of equation (3.9) may have acquired their protons either from the acid added to the solution or from the dissociation of water. In the latter situation, every proton supplied yields a residual hydroxyl ion.) Thus

$$[HA] = H_t - [H] + K_w[H]^{-1} = A_t - [A] \tag{3.10}$$

and measurement of [H] leads to values of [HA] and [A] provided that H_t and A_t are known and that $K_w[H]^{-1} \ll 1$ (or that K_w has been measured under identical conditions). Values of K_{diss} and β then follow.

This all looks very easy, particularly for acidic solutions (pH < 7) where the term $K_w[H]^{-1}$ can be neglected. But general questions arise. Can potentiometric measurements really yield values of [H]; or do they rather give {H}; or neither? How must the procedure outlined above be modified if values of {H}, rather than of [H], are measured? And how should values of $_aK$ and $_BK$ be obtained? What experimental procedure provides a convenient way of varying A_t, or H_t, or both? How may values of the constants best be obtained from the measurements? What simplifications have taken place in the procedure in the course of its transformation into the form encountered in elementary textbooks, and when are they justified? How, if at all, can the treatment be extended to embrace multistep protonation and yet more sophisticated equilibria? We shall attempt to answer all these questions in turn.

Potentiometric determination of [H]

This can indeed be carried out, using the cell:

| $-$ reference half-cell | salt bridge | test solution in constant medium | probe for H$^+$ | $+$ | (3.11) |

provided that the concentration of H$^+$ is low compared with that of the background cation. The cell is first calibrated in the same ionic medium by titrating a volume V_a of strong acid of known concentration H_i with titres v_b of strong base of known concentration B_i. At any point in the titration, the value of

$$H_t = [H] - [OH] = (V_aH_i - v_bB_i)/(V_a + v_b) \qquad (3.12)$$

can be calculated.

Now from the Nernst equation, the e.m.f. of cell (3.11) is given by

$$E = \text{constant} + E_j + RTF^{-1} \ln \{H\} \qquad (3.13)$$

where E_j is the diffusion potential at the junction j.
Since, before the equivalence point, $H_t = [H]$, the value of

$$E^{0\prime} = E - RTF^{-1} \ln [H] = \text{constant} + RTF^{-1} \ln \gamma_H + E_j \qquad (3.14)$$

may be calculated. Since a constant value of $E^{0\prime}$ $(=E^0)$ implies that $(RTF^{-1} \ln \gamma_H + E_j)$ is also constant, E^0 can be used as a calibration constant for calculating [H] from the experimental value of E within that range over which the calibration was performed. It is often found that the value of $E^{0\prime}$ is indeed constant in media where H^+ ions form only a small proportion of the total cations. In more acidic solutions, it is commonly found that

$$E^{0\prime} = C + C'[H] \qquad (3.15)$$

where C and C' are constants. Once C and C' have been determined, equations (3.14) and (3.15) can be used to calculate [H] from E by successive approximation.

Potentiometric determination of {H}

Measurement of 'pH' by means of a commercial pH-meter involves comparing the e.m.f. E_s of the cell:

$$-\text{calomel half-cell} | \text{KCl bridge} | \text{test solution} | \text{glass electrode} + \qquad (3.16)$$

containing a standard buffer with the e.m.f., E, of the analogous cell containing a solution of unknown pH, defined by

$$pH = pH_{standard} + \frac{E_s - E}{RTF^{-1} \ln 10} \qquad (3.17)$$

It is clear from the Nernst equation (3.13) that

$$pH - pH_{standard} = \log \{H\}_{standard} - \log \{H\} \qquad (3.18)$$

if, and only if, there is no appreciable difference between the junction potential in the cell containing the test solution and that of the cell containing a standard buffer. This condition will be fulfilled if the test solution and the standard do not differ too widely in pH. Since the value of $\{H\}_{standard}$ has been fixed by international convention, based on certain explicit assumptions, the value of {H} in the unknown solution may be obtained, subject, of course, to the same assumptions. Special care must, however, be taken when interpreting pH measurements in solutions in mixed aqueous–organic solvents. A similar but more rigorous procedure has been used by the Harned school for obtaining values of {H} from the e.m.f. of cells without liquid junction.

The use of {H} in calculations of constants

The three subspecies of dissociation constants of HA may be written in terms of {H} to give

$$_cK_{diss} = \frac{\{H\}[A]}{\gamma_H[HA]} \tag{3.19}$$

$$_aK_{diss} = \frac{\{H\}[A]\gamma_A}{[HA]\gamma_{HA}} \tag{3.20}$$

$$_BK_{diss} = \frac{\{H\}[A]}{[HA]} \tag{3.21}$$

The mass balance expression (3.10), from which the values of [HA] and [A] are obtained, may be similarly refurbished to give

$$[HA] = A_t - [A] = H_t - \{H\}\gamma_H^{-1} + K_w\gamma_H\{H\}^{-1} \tag{3.22}$$

So rigorous calculation of $_cK_{diss}$ and $_BK_{diss}$ from {H}, H_t and A_t requires knowledge of γ_H, while that of $_aK_{diss}$ involves the activity coefficients of all three species. It may, however, be possible to introduce simplifications, such as the assumption that $[H] \ll H_t$; then $H_t \simeq [HA]$, so that $_BK_{diss}$ can be calculated from the data {H}, H_t, A_t without bothering with activity coefficients.

In the general case where activity coefficients cannot be ignored, we optimistically assume that the activity coefficient of an ion depends only on its charge, and on the ionic strength I of the solution. The ionic strength is given by

$$I = \tfrac{1}{2} \sum z_S^2[S] \tag{3.23}$$

and so depends only on the concentration [S] and charge z_S of each species S. More specific effects, such as the size of an ion, are firmly, if improperly, ignored. In dilute aqueous solutions ($I \leqslant 0 \cdot 1$M) at room temperature, an acceptable value of the activity coefficient of an ion is given by

$$-\log \gamma_S = \frac{\tfrac{1}{2} z_S^2 I^{\frac{1}{2}}}{1 + I^{\frac{1}{2}}} \tag{3.24}$$

The values of $[H] = \{H\}\gamma_H^{-1}$ and γ_H may be obtained from a measured value of {H} by successive approximations. (First assume $\gamma_H = 1$, so that $\{H\} = [H]$ and use this value of [H] to calculate I from equation (3.23), and hence γ_H from equation (3.24). Use this new value of γ_H to refine the values of [H], I and γ_H and repeat until they converge.) The final value of γ_H is fed into equation (3.22) to give values of [HA] and [A]. And, since we now know {H},

γ_H, [HA] and [A] we may calculate values of $_cK_{diss}$ and $_BK_{diss}$ (and of $_c\beta$ and $_B\beta$).

Equation (3.24) implies: (i) that in a given dilute solution, all singly charged ions have the same activity coefficient, γ; and (ii) that the activity coefficients of uncharged species are unity. The method of calculating activity quotients from values of {H}, H_t and A_t therefore depends on whether the species A is an anion or an uncharged molecule. For the equilibrium

$$HA \rightleftharpoons H^+ + A^- \qquad (3.25)$$

we may write

$$_aK = {_cK}\gamma^2 = {_BK}\gamma \qquad (3.26)$$

and calculate values of $_aK$ for a series of measurements at different ionic strengths. Extrapolation of $(_aK)_{calc}$ to infinite dilution eliminates any error caused by the approximations in equation (3.24), and gives a true value of $_aK$. For the dissociation

$$HA^+ \rightleftharpoons H^+ + A \qquad (3.27)$$

of a cationic acid, on the other hand, we may write

$$_aK = {_cK}$$

provided that $\gamma_A = 1$ and $\gamma_H = \gamma_{HA}$. Extrapolation of $_cK$ to infinite dilution eliminates any errors introduced by these assumptions and so gives a value of $_aK$.

Experimental procedure

It is clearly possible to obtain a value of $_BK$ from a measurement of the pH of a single HA–A buffer solution; but this is obviously extremely undesirable. The values of H_t and A_t should be varied over as wide a range as practicable in order that some reliance can be placed on the resultant equilibrium constant, or at least on its limits of error. The upper value of the total concentrations may be fixed by problems of limited solubility or variation of activity coefficients, while the lower limit will be reached when the dissociation equilibria within the medium (usually water, contaminated with carbon dioxide, silica and other undesirables) can no longer be considered negligible compared with the equilibria which we are hoping to study. Despite these limitations, A_t may often be varied by a factor of 10^2, and H_t from, say, 10^{-1}M to -10^{-1}M (where a negative value of H_t implies an excess of strong base). The titration may be carried out in a number of ways. For example, the value of H_t may be varied widely by titrating A with HCl or by titrating HA with NaOH. Both the titrant and the original solution should be prepared in the same ionic medium if the results are to be self-consistent. In each of these titrations, A_t varies only on account of dilution. It is therefore desirable to run several titrations in which H_t is varied for different initial values of A_t. Alternatively, a solution of HCl, or of HA, may be titrated with A. Or H_t and A_t may be varied simultaneously, though not independently, by titrating the medium itself with an HA–A buffer. Again several titrations should be performed, with different ratios of HA to A.

The titration vessel should, of course, be thermostatted. The solution may conveniently be stirred with a stream of nitrogen, pre-saturated with the medium. Some workers prefer to use a magnetic stirrer, but if it becomes undisciplined, it may damage the glass electrode. A typical titration cell, and set of titration curves obtained from it, are shown in Figs 3.1 and 3.2.

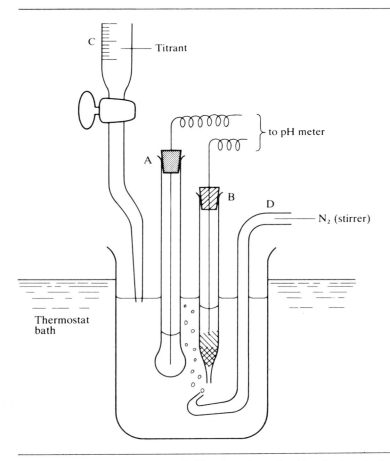

Fig. 3.1 A typical pH-titration cell. The glass electrode (A) and calomel reference electrode (B) are sometimes manufactured as a single unit. The burette (C) may be replaced by a microsyringe and the nitrogen inlet (D) by a magnetic stirrer.

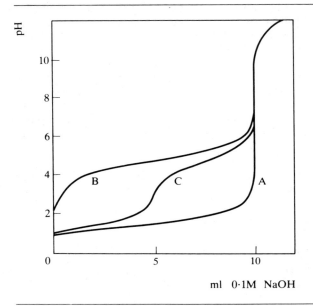

Fig. 3.2   pH-titration curves for the addition of 0·1M NaOH to: (A) 10 ml 0·1M HCl; (B) 10 ml 0·1M CH₃COOH; (C) a mixture of 5 ml of A and 5 ml of B.

Processing the measurements

We have seen that the equilibrium constant for a single-step protonation may, in principle, be obtained from one set of values of H_t, A_t and [H] (or {H}). But we should have no way of knowing if the value of a constant so obtained were seriously inaccurate on account of one or more of diverse possible errors, such as bad analysis of stock solutions, bad preparation of titration solutions, misreading of burette or potentiometer, wrong calibration of electrode assembly, faulty electrodes, miscalculation, wrong chemical premises (activity coefficients not unity, reaction not single step), and so on. Even if an accurate value were obtained, we could only estimate its precision from the precision with which we can measure volumes, concentrations and e.m.f. We should have no experimental check on our estimate. So it is clearly better to collect a generous set of values of H_t, A_t and [H] or {H}, and process them to obtain the 'best' value of the constant.

We shall, for simplicity, assume that reliable sets of values of H_t, A_t and [H] have been obtained, in a constant ionic medium, and that the only problem is that of obtaining the 'best' value of the concentration quotient together with some estimate of its precision. Several types of approach have been used, e.g.:

(i) *Direct computation*   Values of [HA] and [A] are obtained from each set of measurements by means of equation (3.10). A value of K_{diss} or β is then calculated from each point, and the individual values processed by standard statistical methods to give a mean value and a standard deviation.

(ii) *The Henderson–Hasselbalch equation*   It follows from equation (3.4) that

$$\log [HA]/[A] = -\log K_{diss} + \log [H]$$

$$= \log \beta \quad + \log [H] \tag{3.28}$$

so that a plot of $\log [HA]/[A]$ against long $[H]$ is a straight line of unit slope. The required constant can be obtained from the intercept. Limits of error in $\log K_{diss}$ or $\log \beta$ may either be estimated visually or obtained by subjecting the input $\log [HA]/[A]$, $\log [H]$ to a linear least squares treatment.

Equation (3.28) is a manifestation of the Henderson–Hasselbalch equation, which is more often encountered in the form

$$pH = -\log \{H\} = pK + \log \frac{[A]}{[HA]} \tag{3.29}$$

where $pK = -\log {}_B K$. If values of $\{H\}$ are used instead of $[H]$ in the method outlined above, the value of ${}_B K$ may be obtained.

(iii) *Non-linear curve fitting*   We may express the fraction of total A in the form of base as α_0 and that in the form of acid as α_1. Then

$$\alpha_0 = \frac{[A]}{A_t} = \frac{[A]}{[A]+[HA]} = \frac{1}{1+\beta[H]} = \frac{1}{1+K_{diss}^{-1}[H]} \tag{3.30}$$

and

$$\alpha_1 = \frac{[HA]}{A_t} = \frac{[HA]}{[A]+[HA]} = \frac{\beta[H]}{1+\beta[H]} = \frac{K_{diss}^{-1}[H]}{1+K_{diss}^{-1}[H]} \tag{3.31}$$

These relationships indicate that α_0 and α_1 are functions only of $[H]$ and are independent of H_t, A_t or $[A]$. Moreover if we replace $\beta[H] = K_{diss}^{-1}[H]$ by the normalised variable $[\mathbf{H}]$, we see that

$$\alpha_0 = \frac{1}{1+[\mathbf{H}]} \tag{3.32}$$

and

$$\alpha_1 = \frac{[\mathbf{H}]}{1+[\mathbf{H}]} \tag{3.33}$$

The secondary variables α_0 and α_1 are therefore unique functions of $[\mathbf{H}]$ only, and plots of $\alpha_0(\log [\mathbf{H}])$ and $\alpha_1(\log [\mathbf{H}])$ are of unique shape. Since

$$\log [\mathbf{H}] = \log [H] + \log \beta \tag{3.34}$$

$$= \log [H] - \log K_{diss} \tag{3.35}$$

the position of the experimental plots $\alpha_0(\log [H])$ and $\alpha_1(\log [H])$ on the, $\log [H]$ axis depends on the value of the equilibrium constant.

Normalised curves $\alpha_0(\log[\mathbf{H}])$ and $\alpha_1(\log[\mathbf{H}])$ are calculated, very easily, from equations (3.32) and (3.33), plotted on graph paper, and traced together with the axes on to transparent paper. The experimental points α_0, $\log[\mathrm{H}]$ and α_1, $\log[\mathrm{H}]$ are also plotted, on the same scale, but on separate graph paper. The normalised curves are then superimposed on the experimented points in the position of best fit, and the required value of β or K_{diss} obtained from the horizontal displacement of the theoretical and experimental functions (see Fig. 3.3). The acceptable shift from this position indicates

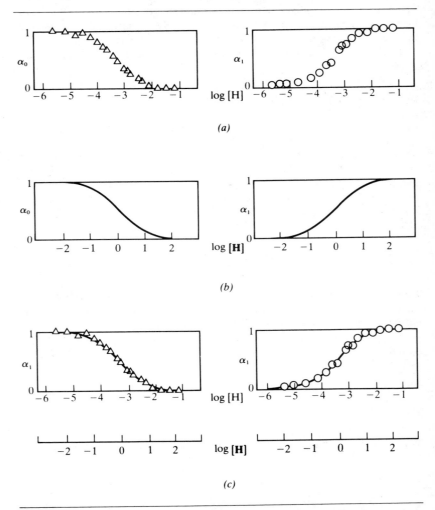

Fig. 3.3   Non-linear curve-fitting. (*a*) experimental points α_c, $\log[\mathrm{H}]$; (*b*) theoretical, normalised curves $\alpha_c(\log[\mathbf{H}])$; (*c*) diagrams *a* and *b* superimposed in the position of best fit, corresponding to p$K = 3\cdot4$.

the precision of the value obtained. The curve-fitting method enables those who are sensitive to visual representations of data to get an immediate 'feel' for the system being studied. The range and precision of the measurements and the effect of both random and systematic errors, can be absorbed at a glance; and the method is particularly suitable for projects involving the study of a number of single-step protonations.

Simplifications

Many of the expressions in this chapter, even if apparently unfamiliar, will have been already encountered in only slightly simplified form. We shall look at the procedure outlined earlier and consider what simplifications have been introduced, and with what justification. We shall start by whittling down equation (3.22).

Many workers, whilst measuring {H} with a conventionally standardised pH-meter, assume that $\gamma_H = 1$ for the purposes of substitution into equation (3.22). This introduces only little error provided that both [H] and [OH] are very small compared with H_t. A more drastic assumption is that, for a buffer solution prepared by adding HA to NaA (or HACl to A), no A reacts with water. Thus, in an acetate buffer, it would follow that $[HA] = H_t$ and $[H] = [Na]$. (Similarly, in a solution containing ammonium chloride and ammonia, we would set $[NH_4] = [Cl] = H_t$.) This accounts for the last term in the Hasselbalch–Henderson equation frequently being written as 'log [salt]/[acid]' where the concentrations refer to initial values for NaA and HA respectively. This approximation is only valid if [H] and [OH] are not only small, but completely negligible in comparison with H_t.

An inexcusable simplification frequently occurs in the calculation of K_{diss} from a pH titration. If, for example, HA is titrated with NaOH, it is often assumed that, when exactly half the equivalent amount of NaOH has been added, the concentrations of HA and A are equal. This assumption is tantamount to setting [A]/[HA] equal to the initial ratio [salt]/[acid] as discussed in the previous paragraph; and provided that the acid is not very strong, very weak or very dilute, it may well be justified. What is clearly not justified is the far too common practice of obtaining the value of K_{diss} from the half-neutralisation point alone. It is true that, at this point, $pH = pK_{diss}$ (see equation (3.28)); but there seems little point in measuring a whole titration curve merely to determine two points; the volume v_e of titre at the equivalence point, and the value of pH corresponding to a titre of $v_e/2$. The 'constant' so obtained is hardly more reliable than that calculated from a single measurement of pH and much less efficient in terms of return for effort used.

Extensions

Most of the equations used here are specifically designed to describe single-step reactions. But a few, like those new buildings which are left with protruding bricks to be keyed to future extensions, can readily be adapted for more sophisticated use. The expressions for α_0 and α_1 (equations (3.30) and (3.31)) are, indeed, special cases of the much more general relationships discussed in sect. 7.3. Terms for additional reaction products may be included as required.

The use of α_0 to denote the ratio $[A]/A_t$, although conventional, is perhaps unfortunate. For a solution containing the monobasic acid HA as sole solute, it so happens that α_0 is identical to the 'degree of dissociation α', a concept which is often deeply imprinted on the youthful mind. It sometimes proves difficult to jettison either the concept or the symbol in favour of a more flexible treatment. It may become an almost reflex response to write $[HA] = (1 - \alpha)c$, an expression which, being true only for a c M solution of pure HA, impedes thought about more complex systems. The shackles of the subscriptless α should therefore be shed as soon, and as completely, as possible.

Of all the equations used so far, the expression (3.31) for α_1 has perhaps the greatest potential. We may not be particularly interested in the fraction of A_t in the form of HA. But the ratio $[HA]/A_t$ also represents the average number of protons which are bound to each base A. This average may readily be obtained experimentally as $(H_t - [H] + [OH])/A_t$ regardless of the complexity of the system. And the concept of the average number of bound groups forms the basis of a large part of equilibrium chemistry.

3.3   Spectrophotometry

We have discussed potentiometric methods for measuring K_{diss} and β at some length because they are both good and widely applicable; but they are not, unfortunately, universally applicable. Some acids or bases are insufficiently soluble to be studied in this way. Others are so strong, or so weak, that $(H_t - [H] + [OH])$ differs appreciably from H_t only at such high concentrations as to thwart any attempts to control, or to make sense of, activity coefficients. Some systems need to be studied in non-aqueous solvents in which pH measurements cannot be made. For difficult cases some spectrophotometric technique may be applicable; and even for more tractable systems an absorptiometric method may be a good second best.

The basis of absorptiometry is the Beer–Lambert law, which relates the absorbency A_s of a solution to the concentrations (not to the activities) of the species present. If monochromatic radiation of wavelength λ and incident intensity I_0 is of intensity I when it has passed through a thickness l of solution, we may write

$$A_s = \log \frac{I_0}{I} = l \sum \varepsilon_s[S] \tag{3.36}$$

where the constant ε_S is the extinction coefficient of the species S at wavelength λ. Many organic acids have markedly different ultraviolet absorption spectra $\varepsilon_s(\lambda)$ from their conjugate bases and, since the values of ε_A or ε_{HA} or both are often high ($\sim 10^3 \, M^{-1} \, cm^{-1}$), measurable absorbencies may be obtained in a 1-cm cell using fairly dilute solutions ($\sim 5 \times 10^{-4} M$).

Some acid–base pairs have very intense, pH-sensitive absorption spectra in the visual region and so may be used as acid–base indicators.

We shall first consider an acid–base pair HA–A in an aqueous solution to which strong acid, strong base, or simple buffers may be added. If HA and A are the only species which absorb radiation of the wavelength used, the Beer–Lambert law takes the form

$$A_s = l(\varepsilon_A[A] + \varepsilon_{HA}[HA]) \tag{3.37}$$

Now if a series of solutions which contain the same total concentration of A_t of A is prepared over a wide range of pH, then

$$\lim A_s \ (pH \to 0) = l\varepsilon_{HA}A_t \tag{3.38}$$

and

$$\lim A_s \ (pH \to \infty) = l\varepsilon_A A_t \tag{3.39}$$

so that

$$\tfrac{1}{2}\{\lim A_s \ (pH \to \infty) + \lim A_s \ (pH \to 0)\} = l\{\varepsilon_A A_t/2 + \varepsilon_{HA}A_t/2\} \tag{3.40}$$

The mean of the absorbencies at the extremes of the pH scale therefore gives the absorbency of the solution in which $[A] = [HA] = A_t/2$. Granted the assumptions on p. 27, we have, for this solution

$$[H] = K_{diss} = \beta^{-1} \quad \text{and} \quad \{H\} = {}_BK_{diss} = {}_B\beta^{-1}$$

Although this procedure is frequently used, and can be carried out at a number of wavelengths for which ε_{HA} differs markedly from ε_A, it is in principle little better than obtaining pK from the point of half-neutralisation of a pH titration (see p. 27). As emphasised above, it is always advisable to make the fullest possible use of the measurements which have been obtained.

We may conveniently define an observed extinction coefficient $\mathscr{E}$ as

$$\mathscr{E} = \frac{A_s}{A_t l} = \frac{\varepsilon_A[A] + \varepsilon_{HA}[HA]}{A_t} \tag{3.41}$$

whence, from equations (3.30) and (3.31)

$$\mathscr{E} = \varepsilon_A \quad \varepsilon_{HA}\alpha_1 \tag{3.42}$$

$$= \frac{\varepsilon_A + \varepsilon_{HA}\beta[H]}{1 + \beta[H]} = \frac{\varepsilon_A + \varepsilon_{HA}K^{-1}[H]}{1 + K^{-1}[H]} \tag{3.43}$$

$$= \frac{\varepsilon_A + \varepsilon_{HA \ B}\beta\{H\}}{1 + {}_B\beta\{H\}} = \frac{\varepsilon_A + \varepsilon_{HA \ B}K^{-1}\{H\}}{1 + {}_BK^{-1}\{H\}} \tag{3.44}$$

Values of the concentration quotients and Brönsted constants may therefore be obtained (see Figs 3.4 and 3.5) from the expression

$$\frac{\mathscr{E} - \varepsilon_A}{\varepsilon_{HA} - \mathscr{E}} = \frac{[HA]}{[A]} = \beta[H] = K^{-1}[H] \tag{3.45}$$

$$= {}_B\beta\{H\} = {}_BK^{-1}\{H\} \tag{3.46}$$

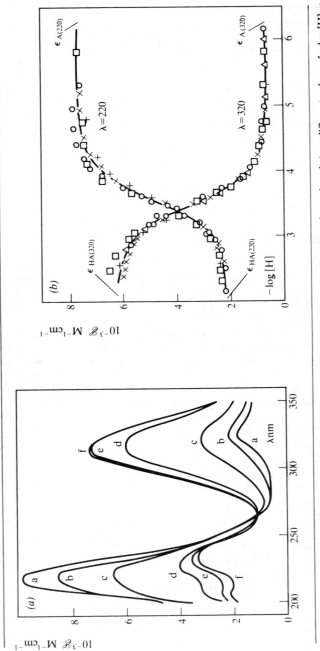

Fig. 3.4   Determination of ε_A and ε_{HA}. (a) Observed extinction coefficient $\mathscr{E}$ as a function of wavelength λ at different values of $-\log[H]$: a, 4·77; b, 4·03; c, 3·67; d, 3·21; e, 3·02; f, 2·86. (b) $\mathscr{E}$ as a function of $-\log[H]$ for various concentrations between 0·0009M and 0·000 45M. (Data for the protonation of the hydrogen rhodizonate ion from D. Alexandersson and N.-G. Vannerberg, *Acta Chem. Scand.*, **26**, 1909 (1972).

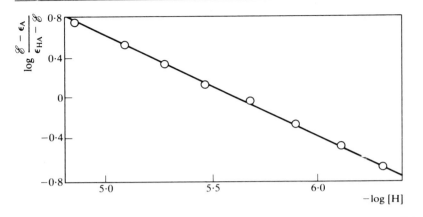

Fig. 3.5 Determination of K for acridine by spectrophotometry. The line corresponds to p$K = 5·62$. (Data from A. Albert and E. P. Serjeant, *The Determination of Ionisation Constants*, Chapman and Hall, London (1971).)

The values of ε_{HA} and ε_A are obtained from the limiting values of $\mathscr{E}$ in solutions of very low and very high pH. The value of [H] or {H} is obtained potentiometrically, as outlined in the previous section. So the method involves two types of instrumental measurement (of A_s and of [H] or {H}), and knowledge of A_t from the overall composition of the solution. Analysis of the measurements involves determination of the two parameters ε_A and ε_{HA} in addition to the required equilibrium constant. (By contrast, potentiometric determination requires a knowledge of H_t in addition to A_t, but uses only one type of measurement and involves no additional parameters.)

An early ultraviolet spectrophotometric study of dissociation was carried out in solutions of varying A_t containing background salt but no buffer. The value of ε_A was obtained from solutions of NaA, and that of HA from solutions which were sufficiently concentrated for dissociation to be negligible. Values of $[A] = [H] = \alpha_0 A_t$ and $[HA]\alpha = \alpha_1 A_t = (1-\alpha_0)A_t$ were obtained from the absorbency by means of equation (3.42) and no separate measurement of [H] or {H} was needed. Somewhat similar procedures have been used to study strong acids by more recent spectroscopic techniques. For example, Raman intensities measured at 1049 cm^{-1} in aqueous NaNO$_3$ and HNO$_3$ have been used to calculate values of α_0 in aqueous nitric acid of concentrations up to 16M. Sulphuric and iodic acids have been similarly studied. But it is, of course, difficult to relate values of α_0 with those of $_BK$ or $_aK$ for systems in which there are such gross variations in activity coefficients as would be expected between, say, dilute and 16M nitric acid.

Strong acids have also been studied by proton magnetic resonance. Since exchange between H$_3$O$^+$ and HA is rapid ($<10^{-3}$ s) a single line is obtained. The position of this line depends on the values of α_0 and α_1 in a manner formally identical to equation (3.42). Measurements of α_0 so obtained for nitric acid are in satisfactory agreement with those determined by Raman spectroscopy, though the difficulty in interpreting the values of course remains. However, if an acid is strong, evidence for incomplete dissociation must be sought in at least moderately concentrated solution, even though no thermodynamicist of sensitivity would choose to make measurements under these conditions. Recalcitrant systems do, however, require sledge-hammer treatment and before we superciliously condemn the work as thermodynamically non-rigorous we should decide whether we prefer non-ideal information, or none at all.

A different application of spectrophotometry to equilibrium chemistry is the use of indicators. If the acid–base pair being studied is transparent, the value of pH (or of a related quantity) may be determined by adding an indicator pair HIn–In for which the values of one or both extinction coefficients ε_{HIn} and ε_{In} are high. The ratio [HIn]/[In] may be measured spectrophotometrically, using the analogue of equation (3.45), and combined with known values of K_{diss} (or $_BK_{diss}$) for HIn to give [H] (or {H}). The required values of [HA] and [A] may be obtained, by successive approximation if necessary, from the mass balance relationships

$$[HA] = H_t - [H] + [OH] - In_t(1 + [In]/[HIn])^{-1} \tag{3.47}$$

and

$$[A] = A_t - [HA]$$

The required value of K_{diss} (or $_BK_{diss}$) for HA can then be calculated.

Those who work with activity quotients have suggested various ways of dealing with our ignorance of activity coefficients of HIn, In, HA and A. But we shall not discuss them as the method is inferior to potentiometry for studying aqueous solutions. It has, however, been widely used in non-aqueous solvents in which glass electrodes behave very erratically. Hammett defined an 'acidity function'

$$H_0 = -\log\{H\}\gamma_{In}/\gamma_{HIn} = -\log {}_aK_{diss} - \log \frac{[HIn]}{[In]} \tag{3.48}$$

which has been much used in spectrophotometric studies of the relative strengths of very weak bases in acidic solvents. Use of the Hammett acidity function should be limited to solvents of high dielectric constant; in those of low dielectric constant, such as glacial acetic acid, more than the proton may join with a base.

3.4 Distribution methods

The charged member of the HA–A pair can normally exist only in an ionising solvent, but the uncharged member may often be able to partition between the ionising solvent and a second phase in contact with it. If the equilibrium constant for this partition can be determined, measurement of the total concentration of A in each phase leads to the fraction, α_u, of the total A in the aqueous solution which is present as the uncharged species. So if we measure α_u as a function of [H] in the aqueous phase, we can calculate K_{diss} and β from equations (3.30) to (3.33). Distribution methods have not been widely used and so will be discussed only briefly. They may, however, be useful for systems where the uncharged species is only sparingly soluble in water.

We shall consider first the partition of an uncharged acid HA between water and an organic solvent which is effectively immiscible with it. If all of the A in the organic phase is present as HA, the ratio, q, of the total concentrations of A in each phase is given by

$$q = \frac{(A_t)_{org}}{(A_t)_{aq}} = \frac{P_1[HA]_{aq}}{(A_t)_{aq}} = P_1\alpha_1 \tag{3.49}$$

where $P_1 = [HA]_{org}/[HA]_{aq}$ is the partition coefficient of HA. The analogous expression for the partition of an uncharged base A is

$$q = \frac{(A_t)_{org}}{(A_t)_{aq}} = \frac{P_0[A]_{aq}}{(A_t)_{aq}} = P_0\alpha_0 \tag{3.50}$$

where $P_0 = [A]_{org}/[A]_{aq}$. The pH of the aqueous phase should be varied over a wide range and it is convenient to use a series of buffers as well as dilute acid and alkali. The distribution ratio, q, is obtained from analytical determination of the total A in each phase, and the value of [H] or {H} in the aqueous phase is measured potentiometrically after equilibration. Extrapolation of q to very high (or very low) values of [H] gives P_1 (or P_0). The required equilibrium constants can then be obtained from the variation of α_u with [H] or {H}. The method has been used for uncharged bases such as aniline, and acids such as 8-hydroxyquinoline. (But equation (3.49) is not applicable to acetic acid, which forms dimers in organic solvents.)

A similar method involves the partition of a sparingly soluble acid or base between the solid and a saturated solution. The concentration, S_1, of the uncharged acid, or S_0, of the uncharged base, will be constant in a saturated solution, but the overall solubility S will, of course, vary with pH. Since $S = A_t$, we may write

$$\frac{S_1}{S} = \alpha_1 \tag{3.51}$$

for an uncharged acid and

$$\frac{S_0}{S} = \alpha_0 \tag{3.52}$$

for an uncharged base. The solubility is measured over a wide range of pH by a standard analytical method, and the value of [H] or {H} in the saturated solution is measured potentiometrically. The value of S_1 or S_0 is obtained by extrapolation of S to very low or very high values of pH.

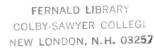

3.5   Electrical conductivity

Before the mid-1930s, almost all reliable values of acid dissociation constants had been obtained by measuring the electrical conductivity of solutions of H^+A^-. Since conductivity depends on the concentrations and mobilities of all charged species, a background electrolyte obviously cannot be used. Nor can values of H_t and A_t be varied independently. The activity coefficients vary, of course, with ionic strength which depends on the extent of dissociation. Even worse, the mobilities of the ions vary with the ionic environment in a way which depends on the nature of the ions as well as on their concentration.

Schoolday encounters with conductivity sensibly ignore all these complications, and we usually first meet the method in the form of the approximate expression

$$\alpha_0 = \frac{\lambda_{\text{solution}}}{\lambda_{(H^+ + A^-)}} \qquad (3.53)$$

The value of the numerator is easily obtained from measurement of the conductivity of the acid, and that of the denominator from similar measurements for separate solutions of HCl, NaCl and NaA. Since $\lambda_{(H^+ + A^-)}$ is shorthand for (HCl + NaA − NaCl), it represents the hypothetical value of $\lambda_{\text{solution}}$ if the solution were completely dissociated. Much ingenuity has been invested in modifying equation (3.53) to allow for the effect of ionic interaction on activity coefficients and mobilities. Given the skill and patience to perform the necessary extrapolations and iterative procedures, the method can yield dissociation constants which would survive the most rigorous of critics; but it is so cumbersome, and so limited, that we shall not discuss it further. The method has, indeed, been described frequently, often in great detail.

3.6   Flashback

We have now at least mentioned almost all the methods by which dissociation constants have been measured during this century. In terms of ease, precision and range of applicability some techniques are clearly more desirable than others. Potentiometry is, surely deservedly, the most popular since it can be adapted to give either the activity or the concentration of the single species H^+, and the parameter E^0 which connects $\{H^+\}$ or $[H^+]$ with the measured values of E can be determined independently using solutions which do not contain the base A.

Liquid–liquid partition can also give the concentration or activity of a single species, in this case the uncharged member of the pair HA–A. The parameter which connects such quantities with the measured distribution ratio is the partition coefficient, which can sometimes, although not always, be determined independently of the acid–base equilibrium. In the solubility method, the constant concentration or activity of the uncharged species is the parameter which enables the experimental measurements (in this case of total A) to be related to the equilibrium concentrations required for calculation of dissociation constants. In favourable cases, this parameter too may be measured independently of the protonation reaction.

Spectrophotometric measurements of solutions in which both HA and A absorb are more difficult to interpret as the absorbency depends on the sum of the products of concentration and extinction coefficient for the two species. But the extinction coefficients are at least true parameters in that they are unaffected by the extent of acid dissociation, and are even (normally taken to be) independent of the ionic medium.

Conductivity is by far the least satisfactory of the methods mentioned so far. Activity coefficients cannot be controlled, and the ionic mobilities, unlike the connecting parameters involved in other methods, vary with the composition of the solution.

In general, we can express a measured property of a solution containing several species S as

$$M = f \sum (q_S[S]) \qquad (3.54)$$

Interpretation is easiest when f is a simple function and when the quantities q_S are zero for all species but one, and the remaining quantity q_S is a constant. Although more difficult, interpretation is quite possible when values of q_S are not zero, provided that they are constant. It is when values of q_S vary within a set of measurements that the real trouble starts.

Chapter 4

Two-step protonation

As the book proceeds, increasingly complex equilibria will be discussed. But since simplifications are often imposed on studies of multiple equilibria, increase in the complexity of the reaction by no means implies a corresponding increase in the difficulty of the treatment. In this chapter, we shall discuss the addition of two protons to a base, and we shall introduce two compensating simplifications.

1. We shall assume here, and throughout the rest of the book, that measurements are made in a medium which contains so large an excess of bulk electrolyte that activity coefficients are constant. We shall define the standard state as the hypothetical molar state in that medium. All activity coefficients are then, by definition, unity, and all equilibrium constants are concentration quotients which will be written without the left-hand subscript $_c$. Those unfortunates who are considering measurements made in the absence of bulk electrolyte will have to allow for variation of activity coefficients and so must combine the present treatment of a dibasic acid in a constant ionic medium with that of a monobasic acid in a variable medium (see Ch. 3). The process, though tedious, is not conceptually difficult. But since it involves 'estimation' of activity coefficients, the use of a bulk electrolyte is strongly recommended, as much to eliminate guesswork as to lighten labour.

2. Our second simplification involves writing chemical reactions as processes in which a complex species is formed by association of two simpler ones, rather than as a charge involving dissociation. We shall therefore discuss the protonation of bases, rather than the dissociation of acids. This makes for simpler typesetting and hence for easier reading and, maybe, even for readier understanding.

4.1 Basic algebra

The treatment which follows describes the addition of H^+ ions to a base A which has two identical sites for protonation; diprotic bases with non-identical sites are discussed in sect. 4.6.

Diprotic bases may differ widely in nature, charge and size. Examples are the oxalate, malonate, sulphate, selenide and hydroxyl ions, and the molecule ethylenediamine. The first protonation step is obviously

$$H + A \rightleftharpoons HA \tag{4.1}$$

But the formation of the second acid H_2A may be represented either by the single-step protonation of HA

$$HA + H \rightleftharpoons H_2A \tag{4.2}$$

or by the overall process

$$A + 2H \rightleftharpoons H_2A \tag{4.3}$$

There are therefore two equilibrium constants for the formation of the doubly protonated species: the stepwise formation constant, represented by

$$K_2 = \frac{[H_2A]}{[HA][A]} \tag{4.4}$$

and the overall formation constants, written as

$$\beta_2 = \frac{[H_2A]}{[H]^2[A]} \tag{4.5}$$

The formation constant for the first acid, HA, can be written as either K_1 or β_1 as stylistic considerations suggest. So

$$K_1 = \beta_1 = \frac{[HA]}{[H][A]} \tag{4.6}$$

whence

$$\beta_2 = K_1 K_2 \tag{4.7}$$

or

$$K_2 = \frac{\beta_2}{\beta_1} \tag{4.8}$$

Reverting briefly to acid dissociation constants, we note that first and second constants refer, respectively, to the loss of the first and second protons from the dibasic acid H_2A. They are thus the equilibrium constants for the two stepwise dissociations

$$H_2A \rightleftharpoons HA + H \tag{4.9}$$

and

$$HA \rightleftharpoons A + H \tag{4.10}$$

since for the first dissociation (4.9),

$$K_{\text{diss(1)}} = \frac{[HA][A]}{[H_2A]} = K_2^{-1} \tag{4.11}$$

and for the second, (4.10),

$$K_{\text{diss}(2)} = \frac{[\text{HA}][\text{A}]}{[\text{HA}]} = K_1^{-1} \qquad (4.12)$$

the first dissociation constant is the reciprocal of the second stepwise formation constant, and vice versa. Thus

$$pK_{\text{diss}(1)} = \log K_2 \quad \text{and} \quad pK_{\text{diss}(2)} = \log K_1$$

These confusing relationships should always be borne in mind when using data culled from the literature. Here we shall restrict ourselves to formation constants (stepwise, or overall, as convenient). The equations which follow may, however, readily be recast in terms of $K_{\text{diss}(1)}$ and $K_{\text{diss}(2)}$ if so required.

The treatment of equilibria involving two protonation steps is exactly analogous to that described in Chapter 3 for the addition of a single proton to a base. It is again possible to express the total concentrations, A_t and H_t, of A and of dissociable hydrogen in terms of the equilibrium concentration of just one of the species present together with one parameter for each protonation step involved. For aqueous solutions containing the species H, OH, A, HA and H_2A a maximum of three parameters is needed: β_1, β_2 and (unless [OH] is negligible) K_w. Thus

$$A_t = [\text{A}] + [\text{HA}] + [\text{H}_2\text{A}] \qquad (4.13)$$

and

$$H_t = [\text{H}] + [\text{HA}] + 2[\text{H}_2\text{A}] - [\text{OH}] \qquad (4.14)$$

cf. equations (3.8) and (3.9). The free concentrations of HA, H_2A and OH may conveniently be replaced by the equilibrium concentration variables [H] and [A] together with the appropriate equilibrium constants. The symbols β_1 and β_2 give briefer equations tha K_1 and K_2, so we may write

$$A_t = [\text{A}] + \beta_1[\text{H}][\text{A}] + \beta_2[\text{H}]^2[\text{A}] \qquad (4.15)$$

and

$$H_t = [\text{H}] - K_w[\text{H}]^{-1} + \beta_1[\text{H}][\text{A}] + 2\beta_2[\text{H}]^2[\text{A}] \qquad (4.16)$$

One of the two concentration variables may now be eliminated by combining equations (4.15) and (4.16). Since the value of [H] may normally be measured potentiometrically, it is usual to eliminate [A] by substitution of

$$[\text{A}] = \frac{A_t}{1 + \beta_1[\text{H}] + \beta_2[\text{H}]^2} \qquad (4.17)$$

into equation (4.16). We then obtain the very useful expression

$$\frac{H_t - [\text{H}] + K_w[\text{H}]^{-1}}{A_t} = \bar{j} = \frac{\beta_1[\text{H}] + 2\beta_2[\text{H}]^2}{1 + \beta_1[\text{H}] + \beta_2[\text{H}]^2} \qquad (4.18)$$

The left-hand side of equation (4.18), which can be calculated from experimental values of A_t, H_t and [H] (together with K_w if the pH is high), is already familiar (sect. 3.2) as the average number of protons bound to each base. For a solution which contains a number of acids H_jA (e.g. $j = 0$ for A, $j = 1$ for HA and $j = 2$ for H_2A), this quantity represents the average value of j, and is accordingly denoted by $\bar{j}$. Equation (4.18) shows that $\bar{j}$ is a function only of [H] and is independent of A_t and H_t. The way in which $\bar{j}$ varies with [H] depends on the values of β_1 and β_2 and is best illustrated by considering the shapes of plots of $\bar{j}$ against log [H], for a number of bases which can add two protons.

4.2 Formation curves

Several 'formation curves', as the plots $\bar{j}(\log [H])$ are called, are shown in Fig. 4.1. They all start at a plateau value of $\bar{j} = 0$ at the lowest values of [H]

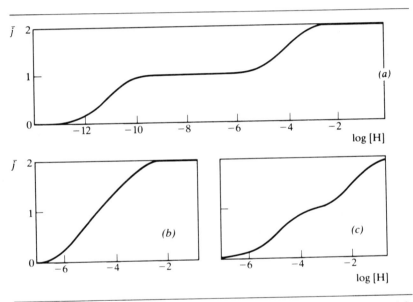

Fig. 4.1 Formation curves $\bar{j}(\log [H])$. (*a*) Dihydrogen selenide, $\log K_1 = 11.0$, $\log K_2 = 3.89$. (*b*) Adipic acid, $\log K_1 = 5.41$, $\log K_2 = 4.41$. (*c*) Oxalic acid, $\log K_1 = 4.28$, $\log K_2 = 1.25$. (From H. Rossotti, *Chemical Applications of Potentiometry*, Van Nostrand, London (1968).)

and reach a plateau of $\bar{j} = 2$ at high acidities. These terminal flat regions obviously describe solutions in which the base A is either completely unprotonated or is protonated to saturation. The central part of the plot may be a single sigmoid curve which rises almost linearly between about $\bar{j} = 0.5$ and $\bar{j} = 1.5$ as in Fig. 4.1(*b*). Some formation curves, however, are composed of two distinct sigmoid steps, separated by a long plateau at $\bar{j} = 1$ (cf. Fig. 4.1(*a*)). Others, such as that in Fig. 4.1(*c*), are of an intermediate, wiggly shape. But whatever the exact shape of the plot, formation curve for a dibasic acid is always symmetrical about its mid-point; the plot for the region $1 \leqslant \bar{j} \leqslant 2$ is identical to that for the region $0 \leqslant \bar{j} \leqslant 1$, rotated through 180°. The plots for different systems differ, of course, in their position on the log [H] axis as well as in their shape. The stronger the base, the lower the concentration of hydrogen ions at which protonation occurs.

For a more quantitative discussion of formation curves, we may rearrange equation (4.18) to give the two apparently dissimilar relationships

$$\bar{j} = \frac{1 + 2K_2[H]}{K_1^{-1}[H]^{-1} + 1 + K_2[H]} \tag{4.19}$$

and

$$\bar{j} + (\bar{j} - 1)\beta_1[H] + (\bar{j} - 2)\beta_2[H]^2 = 0 \tag{4.20}$$

The 'position' of the formation curve may be defined by the value of [H] at the mid-point. Substitution of $\bar{j} = 1$ into equation (4.20) gives

$$[H]_{\bar{j}=1}^{-2} = \beta_2 \tag{4.21}$$

Whilst this expression is true for the formation curve of any dibasic acid, equation (4.19) suggests that the value of $\bar{j}$ will be approximately unity over a range of values of [H] provided that $(K_1[H])^{-1} \ll 1 \gg K_2[H]$. This condition is fulfilled for systems in which $K_1 \gg K_2$ in the region of acidity where $K_1 \gg [H] \gg K_2$, i.e. those for which the protonation steps occur at widely separated regions of [H] at acidities where the base A has been totally transformed into HA, but none of the dibasic acid H_2A has been formed. For a plateau to be obtained at $\bar{j} = 1$, the ratio K_1/K_2 should be at least 10^4. The two protonation steps may then be treated completely separately. In the region of [H] where $1 \gg K_2[H]$, the concentration of H_2A may be ignored. Equation (4.18) then reduces to the relationship (3.31) for a monobasic acid, and may be treated as described in Chapter 3. Measurements in the range $1 \leqslant \bar{j} \leqslant 2$ may be analogously interpreted as the addition of a single proton to the fully-formed acid HA. The average number of protons added to each HA may be defined as

$$\bar{j}_{HA} = \bar{j} - 1 = \frac{\beta_1[H] + 2\beta_2[H]^2}{1 + \beta_1[H] + \beta_2[H]^2} - 1 \tag{4.22}$$

Since the free concentration of A is negligible when $1 \leqslant \bar{j} \leqslant 2$, we may write $[A] < \beta_1[H][A]$, or $1 < \beta_1[H]$. Equation (4.22) then becomes

$$\bar{j}_{HA} = \frac{\beta_2[H]^2}{\beta_1[H] + \beta_2[H]^2} = \frac{K_2[H]}{1 + K_2[H]} \tag{4.23}$$

which is identical to the function $\bar{j}([H])$ for a monobasic acid of formation constant K_2. The formation of a dibasic acid for which $K_1/K_2 > 10^4$ can therefore be treated by considering separately the two pairs of species (A, HA) and (HA, H_2A) by means of the methods previously described for monobasic acids. At no value of [H] will appreciable concentrations of A and H_2A coexist.

We can easily test whether or not the mid-point kink of a formation curve is sufficiently marked to allow us to treat the two protonation steps independently. When $K_1/K_2 > 10^4$, the values of [H] at $\bar{j} = \frac{1}{2}$ and $\bar{j}_{HA} = \frac{1}{2}$ (i.e. $\bar{j} = \frac{3}{2}$) are given by $[H]_{\frac{1}{2}} = K_1^{-1}$ and $[H]_{\frac{3}{2}} = K_2^{-1}$ respectively. Thus if the points $\bar{j} = \frac{1}{2}$ and $\bar{j} = \frac{3}{2}$ are separated by more than 4 units on the log [H] axis, the values of $\log K_1$ and $\log K_2$ are also separated by more than 4 units, and the two protonation steps may be treated independently.

For values of $K_1/K_2 < 10^4$, all three species A, HA and H_2A may coexist in appreciable concentrations. A value of $\bar{j} = 1$ implies that the concentrations of A and H_2A are equal, but not necessarily negligible. For example, when $K_1/K_2 \sim 10$, the concentration of A, HA and H_2A at the mid-point of the formation curve are in the ratio of $19\cdot4 : 61\cdot2 : 19\cdot4$. This is the situation for adipic acid (see Fig. 4.1(*b*)). Even when K_1/K_2 is as high as 10^3, as in the case of oxalic acid (see Fig. 4.1(*c*)), only 94 per cent of the oxalate is in the form of HA when $\bar{j} = 1$; and, because of the symmetry of the curve, the ratio $[HA]/A_t$ is maximal at this point (cf. sect. 13.2).

4.3 Alpha functions

The fraction of A_t in the form of a particular species, A, HA or H_2A may be denoted as α_0, α_1 and α_2 respectively (cf. equations (3.30) and (3.31)). Thus

$$\alpha_0 = \frac{[A]}{A_t} = \frac{[A]}{[A] + [HA] + [H_2A]} = \frac{1}{1 + \beta_1[H] + \beta_2[H]^2} \qquad (4.24)$$

$$\alpha_1 = \frac{[HA]}{A_t} = \frac{\beta_1[H]}{1 + \beta_1[H] + \beta_2[H]^2} \qquad (4.25)$$

$$\alpha_2 = \frac{[H_2A]}{A_t} = \frac{\beta_2[H]^2}{1 + \beta_1[H] + \beta_2[H]^2} \qquad (4.26)$$

and all three quantities α_c, like $\bar{j}$, depend on the single variable [H] and the two parameters β_1 and β_2. Indeed, comparison of equations (4.24), (4.25) and (4.26) shows that

$$\bar{j} = \alpha_1 + 2\alpha_2 = \sum c\alpha_c \qquad (4.27)$$

At the mid-point of the formation curve, α_1 is maximal and $\alpha_0 = \alpha_2$; but $\alpha_1 = 1$ only when $\bar{j} = 1$ and $K_1 \gg K_2$ (cf. sect. 13.2).

4.4 Experimental techniques

Since dibasic acids for which $K_1/K_2 > 10^4$ may be treated by the methods described in Chapter 3, we shall in this chapter mainly discuss ways of studying systems where the protonation steps overlap.

Potentiometric measurement of [H] in a solution of known A_t and H_t gives $\bar{j}$ (cf. equation (4.18)). As is the case for monobasic acids, a series of values of $\bar{j}$, [H], A_t and H_t may conveniently be obtained from a number of potentiometric titrations. If the experimental formation curve $\bar{j}(\log[H])$ looks 'respectable', the measurements may be processed as described below to yield values of the formation constants. Criteria for respectability for the formation curve of a dibasic acid are that:

(a) the points $\bar{j}$, $\log[H]$ form a single curve for all values of H_t and A_t.
(b) the formation curve is horizontal in the regions $\bar{j} \to 0$ and $\bar{j} \to 2$.
(c) the only other permissible plateau or point of inflection is at $\bar{j} = 1$. (And if a plateau does occur at $\bar{j} = 1$ each of the halves of the curve, for $0 \leqslant \bar{j} \leqslant 1$ and $1 \leqslant \bar{j} \leqslant 2$, are of the unique shape defined by equation (3.31) for a monobasic acid.)
(d) the curve is symmetrical about the mid-point where $\bar{j} = 1$.

Any non-random deviation from these criteria implies that something is wrong either with the measurements or with the equilibria postulated to explain them. Thus there may be systematic errors arising from an error in A_t or H_t by faulty analysis of the stock solutions; or some species such as H_3A or HA_2 may be present as well as, or instead of, A, HA and H_2A.

Distribution methods, such as solubility and liquid–liquid partition, may, in principle, be used to study dibasic acids; and even in practice they may be useful for studying acids which are so sparingly soluble that they cannot be studied potentiometrically. If the species H_cA is uncharged, of solubility S_c and of partition coefficient P_c between an immiscible organic solvent and water, the value of α_c may be obtained from the relationships

$$\frac{S_c}{S} = \alpha_c \tag{4.28}$$

and

$$q = P_c\alpha_c \tag{4.29}$$

cf. equations (3.49) to (3.52). The distribution of A between an aqueous buffer and either solid H_cA or an immiscible organic solvent is therefore proportional to the value of α_c; and if the value of [H] in the buffers is measured potentiometrically after equilibration, the functions $S_c^{-1}\alpha_c([H])$ or $P_c\alpha_c([H])$ may be obtained. The parameters S_c and P_c can often be obtained by extrapolation by $[H] \to 0$ for $c = 0$, and to $[H] \to \infty$ for $c = 2$ and so the function $\alpha_c[H]$ may be obtained. When $c = 1$, and the protonation steps overlap, it is difficult to obtain values of S_1 or P_1 independently of the equilibrium constants β_1 and β_2. But, even in this case, measurements of S or q as a function of [H] can be processed to give values of the formation constants, as outlined below.

Absorptiometry is not recommended for dibasic acids when the protonation steps overlap. Since

$$\mathscr{E} = \frac{A_s}{A_t l} = \frac{\varepsilon_0[A] + \varepsilon_1[HA] + \varepsilon_2[H_2A]}{[A] + [HA] + [H_2A]}$$ (4.30)

$$= \frac{\varepsilon_0 + \varepsilon_1\beta_1[H] + \varepsilon_2\beta_2[H]^2}{1 + \beta_1[H] + \beta_2[H]^2}$$ (4.31)

the variation of the observed extinction coefficient $\mathscr{E}$ with [H] depends on no less than five parameters: the two formation constants, and the three extinction coefficients ε_i for the species H_iA. Processing of the data is very tiresome unless some of the extinction coefficients can be determined independently, or other simplifications can be made; and the precision of the emergent parameters is correspondingly low. So the method is best avoided. Conductimetric determination of overlapping protonation constants should certainly be avoided whether the ionic species are A^{2-}, HA^- and H^+; A^-, H_2A^+ and H^+; or HA^+, H_2A^{2+}, H^+ and X^-. As we saw in sect. 3.5, the difficulties are sufficiently daunting when only H^+ and A^- are present and there seems to be no justification for extending the method to even less tractable systems.

4.5   Calculation of the constants

Since the measured quantities $\bar{j}$, S, q and $\mathscr{E}$ are all functions of the single variable [H], the values of β_1 and β_2 may in principle be obtained by the solution of as many simultaneous equations as there are unknown parameters. Thus the two formation constants could be obtained from two pairs of measurement of $\bar{j}$ and [H], whereas five sets of data, $\mathscr{E}$, [H], are needed to give all the parameters ε_0, ε_1, ε_2, β_1 and β_2 which describe the variation of optical absorbency with [H]. In practice, many more than this minimal number of measurements are, or should be, made. The problem is how best to process them.

Of the many methods which have been described, some are bad in that they make use of only a low proportion of the measurements, or even give incorrect solutions to the equations. Several are tedious and inelegant; and others give but poor indication of the precision of the parameters obtained. There remains a number of 'good' methods which make full use of the measurements to give reliable value of the parameters, together with a realistic estimate of their precision. The basic choice between a good graphical method and a good computational one must be decided in the light of available facilities and personal temperament. Ideally, the constants should be obtained by one method of each type; it is no less important to use different methods to process the measurements than to perform two or more independent chemical analyses of stock solutions.

We shall accordingly describe two widely different methods for calculating formation from measurements of $\bar{j}([H])$. Methods for dealing with data $S([H])$, $q([H])$ and $\mathscr{E}([H])$ obtained by distribution or absorptiometric experiments will be described only in outline, since they are required less frequently and are less satisfactory to use.

The projection-strip method

One group of graphical methods for obtaining values of β_1 and β_2 from measurements of $\bar{j}$ and [H] is based on the fact that the shape of the formation curve depends on the ratio $\rho = (K_1/K_2)^{\frac{1}{2}}$ $(=\beta_1/\beta_2^{\frac{1}{2}})$ whereas its position on the log [H] axis depends only on β_2. We can exploit this situation by replacing [H] in equation (4.18) by the normalised variable

$$h = \beta_2^{\frac{1}{2}}[H] \tag{4.32}$$

to give

$$\bar{j} = \frac{(K_1/K_2)^{\frac{1}{2}}h + 2h^2}{1 + (K_1/K_2)^{\frac{1}{2}}h + h^2} = \frac{\rho h + 2h^2}{1 + \rho h + h^2} \tag{4.33}$$

Comparison of equations (4.18) and (4.33) shows that normalisation of [H] shifts the formation curve along the log [H] axis (by a distance equal to $\frac{1}{2}\log \beta_2$) but does not affect its shape. A family of normalised formation curves, calculated for various values of K_1/K_2, is given in Fig. 4.2, which

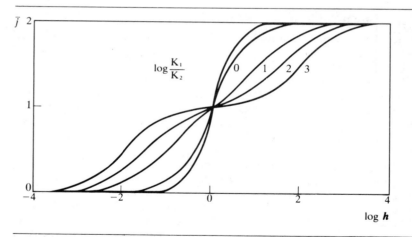

Fig. 4.2 Normalised formation curves $\bar{j}(\log h)$ for different values of the ratio K_1/K_2.

shows that all the curves intersect at the mid-point where $\bar{j} = 1$ and $\log h = 0$. We could, in principle, obtain the value of K_1/K_2 from an experimental curve by calculating theoretical curves with different values of K_1/K_2 until one of identical shape was obtained. The value of $\log \beta_2$ would then be given by the horizontal shift required to make the theoretical and experimental functions coincide. This tedious trial-and-error procedure may be avoided by using the projection-strip method. Instead of using a family of theoretical

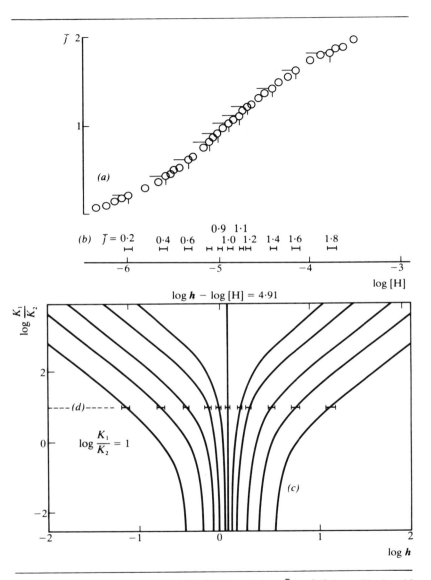

Fig. 4.3 The projection-strip method. (*a*) Typical data $\bar{j}$, log [H] for a dibasic acid (adipic). (*b*) Projection strip of these data for selected values of $\bar{j}$. (*c*) Family of theoretical curves $\log K_1 K_2^{-1}$ (log h) for the same values of $\bar{j}$. (*d*) Strip superimposed on theoretical curves in position of best fit corresponding to $\log \beta_2 = 9\cdot82$ and $\log K_1 = 5\cdot41$.

curves of $\bar{j}$ plotted against log h for various values of K_1/K_2, the projection-strip method makes use of the set of plots of log K_1/K_2 against log h calculated, again from equation (4.33), for a number of constant values of $\bar{j}$ (Fig. 4.3(c)). Since the formation curve for a dibasic acid is symmetrical, the curve for $\bar{j}$ is identical to that for $(2 - \bar{j})$ reflected about the vertical line, log $h = 1$, for $\bar{j} = 1$. Thus, for a given ratio K_1/K_2, the values of log h for $\bar{j}$ and $(2 - \bar{j})$ differ only in sign. The experimental formation curve is plotted with the same horizontal scale as that used for the theoretical curves (Fig. 4.3(a)). The values of log [H] corresponding to each of the values of $\bar{j}$ used in the theoretical curves are marked on the axis, together with estimates of their limits of error (Fig. 4.3(b)). We have now obtained a strip representing a projection of the formation curve on the log [H] axis. This 'projection strip' is superimposed on the theoretical curves, parallel to the log h axis in such a position that, if possible, each point on the strip coincides with the curve calculated for the appropriate value of $\bar{j}$. Failing such perfection, the strip is placed in the position of best fit (Fig. 4.3(d)). The value of log K_1/K_2 is then read off from the ordinate, and that of $\frac{1}{2}$ log β_2 from the difference between the log [H] and log h axes (cf. equation (4.32)). The limits of error in log K_1/K_2 and log β_2 are obtained from the maximum permissible movement of the strip along the vertical and horizontal axes respectively. The projection-strip method is very satisfactory provided that a large number of measurements of $\bar{j}$ and [H] is available. It is easy to use and, once the set of theoretical curves has been plotted, extremely quick. Reliable values of the constants are obtained together with realistic, albeit subjective, limits of error.

Linear plots

A second group of methods by which formation constants of dibasic acids can be obtained from measurements of $\bar{j}$ and [H] involves transforming the experimental data into one of a number of linear functions. The parameters are then obtained either graphically or computationally, from the slope and intercept of the line. For example, yet another manifestation of equation (4.18) is the expression

$$\frac{(\bar{j} - 1)}{\bar{j}}[\text{H}] = \frac{(2 - \bar{j})}{\bar{j}}\frac{\beta_2}{\beta_1}[\text{H}]^2 - \frac{1}{\beta_1} \tag{4.34}$$

A plot of $(\bar{j} - 1)\bar{j}^{-1}[\text{H}]$ against $(2 - \bar{j})\bar{j}^{-1}[\text{H}]^2$ is therefore a straight line of slope K_2 and of intercept $-K_1^{-1}$. These constants may be obtained by eye using a ruler; and this procedure is more convenient than the projection-strip method when values of $\bar{j}$ have been measured over only a narrow range of [H]. But the method really comes into its own as the basis of a sophisticated computational method for obtaining values of the constants and of the appropriate standard deviations by a weighted least squares treatment which makes allowance for the (correlated) experimental errors in the values of both the ordinate and the abscissa (see Ch. 12).

The two methods described above are complementary and, for precise measurements, there is a strong case for using both of them. The visual, projection-strip method has the great advantage that it entails plotting the experimental data as a formation curve which immediately shows up any overall departure from the criteria of respectability (see p. 42) and also any individual 'dud' points (which may then be ignored). Movement of the strip over the theoretical curves allows subjective weighting of data in some ranges of $\bar{j}$ and enables us to see how a change in one or both constants affects the fit. The method involves more personal judgement than does the least squares procedure; but it gives many workers a better 'feel' for a system than columns of values of $(\bar{j}-1)\bar{j}^{-1}[H]$ and $(2-\bar{j})\bar{j}^{-1}[H]^2$ (or even of $\bar{j}$ and $[H]$).

Distribution and spectrophotometric measurements

The variation of solubility, or liquid–liquid partition, with $[H]$ may be described in terms of three parameters: β_1, β_2 and either S_c or P_c. Moreover, the measurements can be represented as a function Q, where

$$Q = p_c \alpha_c \tag{4.35}$$

and p_c is a proportionality constant which depends on some property of the species H_cA. Thus when $Q = q$, then $p_c = P_c$; and when $Q = S^{-1}$, then $p_c = S_c^{-1}$.

The expression (4.35) can also describe spectrophotometric measurements in the special case where only one species, H_cA, absorbs radiation of the wavelength used. Then $Q = \mathscr{E}$ and $p_c = \varepsilon_c$. However, when all three species H_cA absorb, five parameters are involved, and their values may, of course, be obtained by solving sets of five simultaneous equations, preferably by means of a digital computer. But, for a given number of measurements, the precision of each value decreases as the number of parameters increases. Spectrophotometric determination of β_1 and β_2 is therefore best restricted to those acids for which it is difficult to measure $\bar{j}$.

The data Q, $[H]$, like $\bar{j}$, $[H]$, may be processed by visual curve-fitting. The experimental quantities Q, being proportional to α_c, are functions of the single variable $[H]$, which may again be replaced by the normalised variable $h = \beta_2^{\frac{1}{2}}[H]$. For $c = 0$, 1 or 2, we may therefore write

$$\log Q_0 = \log p_0 + \log \frac{1}{1+(K_1/K_2)^{\frac{1}{2}}h + h^2} \tag{4.36}$$

$$\log Q_1 = \log p_1 + \log \frac{(K_1/K_2)^{\frac{1}{2}}h}{1+(K_1/K_2)^{\frac{1}{2}}h + h^2} \tag{4.37}$$

$$\log Q_2 = \log p_2 + \log \frac{h^2}{1+(K_1/K_2)^{\frac{1}{2}}h + h^2} \tag{4.38}$$

So the shape of the plot of $\log Q$ against $\log[H]$ depends only on the ratio K_1/K_2, while its position on the vertical and horizontal axes depends on the values of p_c and β_2 respectively. The three parameters may be obtained by plotting the experimental data as $\log Q$, $\log [H]$ and superimposing them on the family of theoretical curves $\log Q_c (\log \boldsymbol{h})_{K_1/K_2}$, using the same scale throughout. The normalised ordinate $\boldsymbol{Q}_c$ is defined as

$$\boldsymbol{Q} = Q_c p_c^{-1} \tag{4.39}$$

The correct value of K_1/K_2 is found by trial and error as that which gives the theoretical curve of shape identical to that of the experimental plot; and the values of p_c and β_2 are obtained from the experimental coordinates which coincide with the original of the calculated curves. The method is satisfactory if, and only if, Q has been measured over a wide range of [H]; otherwise there may be several equally acceptable positions of 'best' fit.

Various graphical methods have been used to obtain five parameters from spectrophotometric data. They often depend on special relationships between the extinction coefficients (e.g. $\varepsilon_0 = \varepsilon_2$) and use only a low proportion of the measurements. They have little to recommend them except their ingenuity, and the emphasis they throw on the disadvantage of being involved with more parameters than are essential.

The calculated values of β_1 and β_2, and of any other parameters involved, must always be checked by combining them with each experimental value of [H]. The calculated values of $\bar{j}$, S, q or $\mathscr{E}$ so obtained are then compared with the original, experimental values to check that the calculated parameters describe the measurements acceptably over the whole range of [H].

4.6   Some awkward cases

Amino-acids

Throughout this chapter, we have assumed that the base A can add proton to two identical sites; but not all diprotic bases are symmetrical. An amino-carboxylate ion, for example, has two markedly different sites, the uncharged H_2N- group and the negative $-COO^-$ group. Addition of one proton to the glycinate ion, $H_2N-CH_2-COO^-$ could give either the uncharged acid H_2N-CH_2-COOH or the zwitterion $^+H_3N-CH_2-COO^-$, in which the charges are separated. We shall write these species as HA^0 and HA^{+-}. Addition of a second proton to either form gives the same cationic acid $^+H_3N-CH_2-COOH$.

The protonation of an aminocarboxylate ion may be represented by the scheme shown below, there K_1^0 and K_1^+ are the so-called microscopic

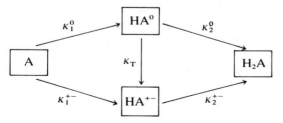

formation constants **of** HA^0 and HA^{+-}, while K_2^0 and K_2^{+-} are the microscopic stepwise formation constants of H_2A **from** HA^0 and HA^{+-} respectively. Since the total concentration of the singly protonated aminocarboxylate is given by

$$[HA] = [HA^0] + [HA^{+-}] \tag{4.40}$$

the value of $K_1 = [HA]/[H][A]$ is given by

$$K_1 = \frac{[HA^0] + [HA^{+-}]}{[H][A]} = K_1^0 + K_1^{+-} \tag{4.41}$$

similarly, we may write

$$K_2 = \frac{[H_2A]}{[H][HA]} = \frac{[H_2A]}{[H]([HA^0] + [HA^{+-}])} \tag{4.42}$$

whence

$$K_2^{-1} = (K_2^0)^{-1} + (K_2^{+-})^{-1} \tag{4.43}$$

Measurements of $\bar{j}$ tell us only about the average number of protons bound to A and not about their position; and so $\bar{j}$ gives values of macroscopic constants K_1 and K_2 rather than those of the four microscopic constants K_1^0, K_1^{+-}, K_2^0 and K_2^{+-}.

From equations (4.41) and (4.42), we see that the tautomeric constant for conversion of HA^0 to HA^{+-} is given by

$$K_T = \frac{[HA^{+-}]}{[HA^0]} = \frac{K_1^{+-}}{K_1^0} = \frac{K_2^0}{K_2^{+-}} \tag{4.44}$$

so that values of the four microscopic constants could be calculated from K_1 and K_2 if a value of either K_T, or of only one microscopic constant, were available. Extremely rough values have been obtained in this way from measurements on esters, together with bold assumptions about the similarity between the behaviour of esters and of (the uncharged form of) the acids from which they are derived. Thus values of K_1^0, K_1^{+-} and K_2^{+-} have been calculated on the assumption that the value of K_1 for an ester $H_2N.(CH_2)_n COOR$ is the same as that of K_2^0 for the acid $H_2N(CH_2)_2COOH$. Similarly, it has been assumed that the optical absorption spectra of these two species are very similar. Then if all of the amino-acid is in one of the monoprotonated forms (i.e. $\bar{j} = 1$ and $K_1 \gg K_2$), the optical absorbency gives $[HA^0]$ and, since $[HA^{+-}] = A_t - [HA^0]$, it also gives the value of K_T. It is not surprising that, given such unrealistic assumptions, the values of the constants obtained vary with the ester used. But a rough value is often better than none, and it is of interest to find that aliphatic aminocarboxylic acids exist predominantly as zwitterions, whereas many of their aromatic analogues favour the uncharged form. Thus for glycine, $K_T \sim 2 \cdot 6 \times 10^5$ while for p-aminobenzoic acid, $K_T \sim 0 \cdot 16$.

If we could observe how the percentage protonation of a particular site varied with pH, we could obtain values of microscopic constants much more satisfactorily, without recourse to values of the macroscopic constants or to dubious assumptions of the types outlined above. Proton magnetic resonance can provide this information. Since the magnetic field acting on any atomic nucleus depends on its detailed chemical environment, the magnetic resonance spectrum of a hydrogen nucleus depends not only on its position in the molecule, but also on whether or not a neighbouring basic site is protonated. This elegant method has been used (Fig. 4.4) to measure the microscopic protonation constants of glutathione and its methylmercury complex.

Carbonic acid

It is well known that the rough pK value of an inorganic oxyacid $(HO)_y YO_z$ can be predicted from the value of z. For most acids with $z = 1$, the pK lies between 2 and 4, but carbonic acid appears to be much weaker than predicted. Potentiometric and conductiometric studies of carbonic acid at equilibrium give $pK \sim 6$.

The discrepancy arises because addition of a proton to the HCO_3^- ion yields an unstable species. Carbonic acid, H_2CO_3, slowly dehydrates to CO_2. So the total concentration of A in solution is given by

$$A_t = CO_3^{2-} + HCO_3^- + H_2CO_3 + CO_2 \tag{4.45}$$

and the observed value of the second stepwise protonation constant is the sum of the two microscopic constants for the reaction

$$H^+ + HCO_3^- \rightarrow H_2CO_3 \tag{4.46}$$

and for the hypothetical step

$$H^+ + HCO_3^- \rightarrow CO_2 + H_2O \tag{4.47}$$

Since H_2CO_3 and CO_2 are solvated, both species are manifestations of H_2A aq, and so cannot be distinguished by equilibrium measurements in the presence of excess of water.

The 'true' value of K_2 for carbonic acid is the equilibrium constant for reaction (4.46), and can be obtained from measurements of electrical conductivity at high field strength. The resistance of solutions of weak electrolytes decreases with increasing field strength, partly because the strong electrical impulse breaks up some of the molecules and partly because the ionic mobilities are increased. So there are more ions and each ion moves faster. The extent of dissociation in a high field is a function of the field strength and of the equilibrium constant at low field. As expected from sect. 4.4, the expression for ionic mobility at high field strength is extremely complicated; but, by using suitable control solutions, equilibrium constants can be coaxed out of measurements of high field conductivity. For studies of carbonic acid, the method has the great attraction that the pulse time ($\sim 4 \mu s$) is so short that the moderately slow desolvation of H_2CO_3 can be neglected. A value of the 'true' K_2 of the system is therefore obtained; and, happily the pK_2 value of $\sim 3 \cdot 9$ is found to fall in the predicted range.

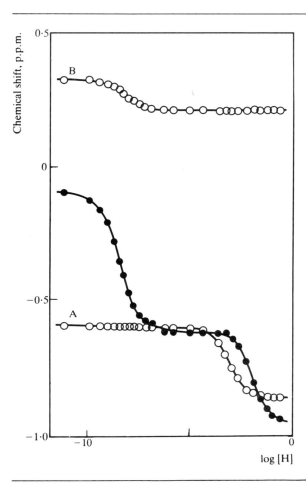

Fig. 4.4 Determination of microscopic protonation constants of glutathione (I) from the pH-dependence of the chemical shift, relative to tetramethyl ammonium nitrate, for protons A, B and C.

$$\underset{\substack{| \\ {}^{+}NH_2}}{\overset{(C)}{HOOC.CH}}.CH_2.CH_2.CO.NH.\underset{\substack{| \\ CH_2SH \\ (B)}}{\overset{(A)}{CH}}.CO.NH.CH_2COOH \qquad (I)$$

(from D. L. Rabenstein, *J. Amer. Chem. Soc.*, **95**, 2797 (1973).)

Chapter 5

Several protonation steps

Equilibrium chemists may be said to count protonation steps using a numerical system akin to those attributed to primitive tribes. According to this unsophisticated system, a base may add one proton, two protons, several protons, or many protons. This chapter deals with bases in the third category, where $3 \leqslant$ 'several' $\leqslant 7$.

We shall again use only *formation* constants (both stepwise and overall) and we must again be very careful when we relate these values to the equilibrium constants (usually given as K_{diss}) in the literature. The correlation between the various equilibrium constants for a tribasic acid H_3A, such as H_3PO_4, as shown below.

Species	Stepwise formation constant	Overall formation constant	Dissociation constant
HA	K_1	$\beta_1\ (=K_1)$	$K_{diss(3)}$
H_2A	K_2	$\beta_2\ (=K_1K_2)$	$K_{diss(2)}$
H_3A	K_3	$\beta_3\ (=K_1K_2K_3)$	$K_{diss(1)}$

The relationship between K_j, β_j and K_{diss} for a species of formula H_jA may be expressed in the general form

$$K_j = \frac{\beta_j}{\beta_{j-1}} = \frac{\beta_1}{K_{diss(J+1-j)}} \tag{5.1}$$

where J is the highest number of protons which can be added to A (i.e. the maximal value of j).

5.1 The shapes of acid formation curves

Luckily, the addition of 'several' protons to a base may take place in very different ranges of pH, in which case the equilibria in each range may be considered separately. Some formation curves are shown in Figs 5.1 and 5.2. The formation curve of phosphoric acid ($K_1/K_2 \sim 10^5$ and $K_2/K_3 \sim 10^5$) shows three distinct steps (Fig. 5.1(a)) which may each be treated by the methods described in Chapter 3 for monobasic acids. The curve (Fig. 5.2(a))

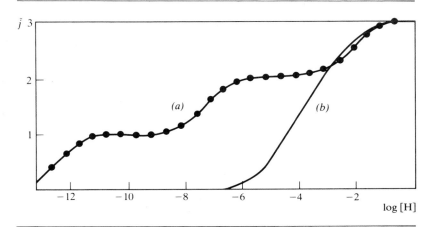

Fig. 5.1   Formation curves for: (*a*) phosphoric, and (*b*) citric acids.

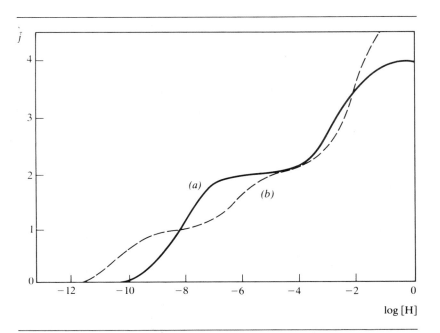

Fig. 5.2   Formation curves for: (*a*) decane-1:10-diphosphoric, and (*b*) ethylenediamine tetra-acetic acids.

for decane-1 : 10-diphosphoric acid, H_4A, shows a marked plateau at $\bar{\jmath} = 2$ (on account of the high value of $K_2K_3 = 10^{4 \cdot 6}$); but there are no plateaux at $\bar{\jmath} = 1$ or $\bar{\jmath} = 3$. The values of $\bar{\jmath}$ in the two regions $0 < \bar{\jmath} < 2$ and $2 < \bar{\jmath} < 4$ may therefore be treated independently by the methods described in Chapter 4 for dibasic acids. The other two acids, shown in Figs 5.1(b) and 5.2(b), are less tractable. The ratios $K_1/K_2 = 10^{1 \cdot 25}$ and $K_2/K_3 = 10^{1 \cdot 4}$ for citric acid are relatively low ($\ll 10^4$) and all four citrate species A, HA, H_2A and H_3A occur in appreciable concentrations in the range of pH $= 4 \cdot 2 \pm 1 \cdot 5$. The ethyl-enediamine tetra-acetate ion can bind up to six protons, the log K_j values being $10 \cdot 25$, $6 \cdot 16$, $2 \cdot 67$, $2 \cdot 21$, $1 \cdot 5$ and ~ 0. The first two protonation steps are therefore widely separated and may be treated by methods suitable for monobasic acids; but the last four steps overlap.

If three or more protonation steps occur over a narrow range of pH, a large number of precise measurements is needed in order to obtain meaningful values of the formation constants. Methods, such as absorptiometry, which introduce yet more unknown parameters, are to be avoided. Following our previous practice of simplifying the discussion in order to compensate for the increasing complexity of the equilibria, we shall deal only with the determination of protonation constants from measurements of $\bar{\jmath}$ and [H].

5.2   Determination of $\bar{\jmath}$

The equations which relate the experimental measurements to the equilibrium constants, via $\bar{\jmath}$, are mere extensions of those which we have already used; and we shall now write them in a more general form. The main difficulty is that of solving the equations to obtain values of the constants. This may be overcome, readily, if a digital computer is available; and, less readily, by determined use of graphical procedures.

The generalised mass balance equations (cf. equations (4.13) and (4.14)) for H and A are given by

$$H_t = [H] + [HA] + 2[H_2A] + \cdots + J[H_JA] - [OH]$$

$$= [H]\left(1 - K_w[H]^{-1} + \sum_0^J j[H_jA]\right) \tag{5.2}$$

and

$$A_t = [A] + [HA] + [H_2A] + \cdots + [H_JA]$$

$$= \sum_0^J [H_jA] \tag{5.3}$$

The values of H_t and A_t are calculated from the analytical composition of the solutions, which are usually prepared by mixing some, or all, of the following ingredients:

	concentration	volume
solution of H_yA	C_1	V_1
solution of B_yA	C_2	V_2
solution of HX	C_3	V_3
solution of BOH	C_4	V_4
solvent		$V_t - (V_1 + V_2 + V_3 + V_4)$

The cation B (e.g. Li^+ or Na^+) should associate with neither A nor OH^-; and the anion X^- (e.g. ClO_4^-) should have no observable basic properties. The 'solvent' is the appropriate ionic medium (usually aqueous BX). The values of H_t and A_t are

$$H_t = (yC_1V_1 + C_3V_3 - C_4V_4)/V_t \tag{5.4}$$

and

$$A_t = (C_1V_1 + C_2V_2)/V_t \tag{5.5}$$

Combination of equations (5.2) and (5.3) gives the general form of the familiar expressions (cf. (3.31) and (4.18)) for $\bar{j}$. As before

$$\bar{j} = \frac{H_t - [H](1 + K_w[H])}{A_t} \tag{5.6}$$

and so

$$\bar{j} = \frac{\sum_0^J j[H_jA]}{\sum_0^J [H_jA]} \tag{5.7}$$

The concentrations $[H_jA]$ of individual acids may again be replaced by the products $\beta_i[H]^i[A]$ to give

$$\bar{j} = \frac{\sum_0^J j\beta_i[H]^i}{\sum_0^J \beta_i[H]^i} \tag{5.8}$$

(Since, for $j = 0$, $[H_jA] = [A]$ and $[H]^i = 1$, the first term on the bottom line of equation (5.8) is unity; and the term for $j = 0$ on the top line is, of course, zero.)

5.3   Calculation of the constants

The J values of β_i for J overlapping equilibria may in principle be obtained by using at least J sets of values of $\bar{j}$ and [H] to solve equation (5.8). If computing facilities permit, all possible combinations of J sets of $\bar{j}$, [H] values may be processed to give the required parameters. Alternatively, values of β_i may be coaxed from the measurements by a multistage, graphical process. Since either method is complicated, the wisdom of controlling activity coefficients, and so dealing only with stoichiometric constants, becomes increasingly apparent the higher the value of J.

Graphical methods for obtaining values of β_i from data $\bar{j}$, [H] are much more involved when $J \geqslant 3$, because formation curves which represent more than two overlapping equilibria have no symmetry properties to be exploited. The least intractable parts of the curve are those to which fewest species contribute to any appreciable extent. These are, of course, the two ends. At low values of [H], the shape of the curve will be determined largely by A, HA and H_2A while at high values, H_JA and $H_{J-1}A$ predominate. So we can concentrate first on the extremes of the curve, and obtain preliminary values of K_1 and K_J, together with rougher preliminary values of K_2 and K_{J-1}. The values of K_1 and K_J may then be used to refine K_2 and K_{J-1} and, in principle to obtain values of K_3 and K_{J-2}. The values of the constants may then be refined by successive approximation. The procedure is described below.

Treatment of data $\bar{j}$, [H] by successive extrapolation

For low values of [H], and hence of $\bar{j}$, we can rearrange equation (5.8) to give

$$\frac{\bar{j}}{(1-\bar{j})[H]} = \beta_1 + \beta_2 \frac{(2-\bar{j})[H]}{(1-\bar{j})} + \beta_3 \frac{(3-\bar{j})[H]^2}{(1-\bar{j})} + \cdots \tag{5.9}$$

which shows that extrapolation of a plot of $\bar{j}/(1-\bar{j})[H]$ against $(2-\bar{j})[H]/(1-\bar{j})$ to zero abscissa (and hence to $[H] \to 0$) gives a value of β_1 as intercept and a rough value of β_2 as limiting slope. The value of β_1 is used to give better value of β_2 with the aid of the expression

$$\frac{\bar{j} - \beta_1(1-\bar{j})[H]}{(2-\bar{j})[H]^2} = \beta_2 + \beta_3 \frac{(3-\bar{j})[H]}{(2-\bar{j})} + \cdots \tag{5.10}$$

The left-hand side is now plotted against $(3-\bar{j})[H]/(2-\bar{j})$ and extrapolated to $[H] \to 0$ to give a value of β_2 as the intercept and a rough value of β_2 as the limiting slope. The process can be repeated *ad lib.*, though since any error in β_1 is carried over to β_2, imprecisions accumulate as the value of j increases.

Values of β_i for high values of j are better obtained from solutions in which [H] is high enough to ensure that the species concerned are present in appreciable concentrations. (But [H] should not be so high that it plays havoc with the junction potential or activity coefficients.) If the two highest acids H_JA and $H_{J-1}A$ are the predominant species, equation (5.8) may be written as

$$\frac{(\bar{j}-J)}{(J-1+\bar{j})}[H] = \frac{\beta_{J-1}}{\beta_J} + \frac{(J-2+\bar{j})}{(J-1+\bar{j})}[H]^{-1} \frac{\beta_{J-2}}{\beta_J} + \cdots \tag{5.11}$$

As before, the left-hand side is plotted against the first variable on the right, and extrapolated to zero abscissa (which in this case corresponds to $[H]^{-1} \to 0$, or $[H] \to \infty$). The intercept before gives K_J^{-1} and the limiting slope gives a rough value of $(K_{J-1}K_J)^{-1}$. The 'stripping' process can be repeated by successive extrapolation to $[H] \to \infty$ in an exactly similar way to that described for extrapolation to $[H] \to 0$. Thus, the next plot, of

$$(\bar{j} - J)[H]^2 - (J - 1 + \bar{j})[H]K_J^{-1}/(J - 2 + \bar{j}) \tag{5.12}$$

against $(J - 3 + \bar{j})/(J - 2 + \bar{j})[H]$, gives β_{J-2}/β_J as the intercept, and β_{J-3}/β_J as the limiting slope; and so on.

We naturally hope that the values of the constants obtained by extrapolation to $[H] \to 0$ are in good agreement with those obtained by extrapolation to high acidity. But when cumulative errors cause discrepancies, the constants should be refined by successive approximation. Suppose that we are uncertain about the value of β_2 for a tribasic acid, but have obtained more acceptable values of β_1 and K_3. Equation (5.8) may be reshuffled to give

$$\beta_2 = \frac{(\bar{j} - 1)\beta_1[H]^{-1} + \bar{j}[H]^{-2}}{(2 - \bar{j}) + (3 - \bar{j})K_3[H]} \tag{5.13}$$

The values of β_2 and β_3 may then be fed back into equation (5.9); as $[H] \to 0$, the last two terms will be small corrections, and a refined value of β_1 can be obtained. This in turn can be used to refine the values of β_2 and β_3 until there is no difference between successive values of β_j for each value of j.

With sufficient patience (and graph paper), 'several' overlapping protonation constants can be obtained in this way; but, as always, the higher the number of parameters which are squeezed out of the measurements, the lower is the precision associated with each of them. The more important it then becomes that the measurements are of the highest quality and that the constants obtained are checked by back-calculation. The observed values of $\bar{j}$ should always be compared with values calculated for the same values of $[H]$ by substituting the supposedly definitive values of the constants into equation (5.8).

Chapter 6

Many protonation steps: the binding of protons to macromolecules

It is seldom possible to coax even as many as six independent protonation constants from measurements of H_t, A_t and [H], however reliable these may be; an increase in the number of species which coexist will clearly cause a decrease in the precision of the parameters which can be extracted from any set of experimental data. Many macromolecules, both natural and synthetic, are acidic, basic or amphoteric; and those macromolecules which interact with protons do so on a vast scale. Ribonuclease, for example, has 36 protonation sites (see Fig. 6.1), haemoglobin has about 130, and poly-methacrylic acid (molecular weight 139 000) has about 1600. The relationships between protons and macromolecules can therefore only be studied with the help of grossly simplifying assumptions. So again, we compensate for the increased complexity of the system by paring down the sophistication of the treatment and thereby decreasing its rigour.

6.1   Identical, independent protonation steps

Studies of equilibria involving protons and macromolecules are usually based on some model of totally unrealistic simplicity. Information about a particular system can then be obtained from the difference between the observed behaviour and that predicted for the model. The traditional starting point is the assumption that all protonation sites are identical and independent. If this assumption were valid, we could describe all the protonation equilibria by means of a single parameter, K, which represents the equilibrium constant for the reaction

$$\text{free site} + \text{H}^+ = \text{protonated site} \tag{6.1}$$

The concentrations of free and protonated site refer to moles of site (rather than of polymer) either in unit volume of solution or, more rigorously, in unit volume of the polymer phase. So, when a macromolecule of total concentration A_t is completely unprotonated, the concentration of free sites is JA_t, where J is the total number of protons which can be bound to each macromolecule. In general, the concentration of protonated sites if $\bar{j}A_t$ and that of free sites is $(J - \bar{j})A_t$. The expression for K is therefore

$$K = \frac{\bar{j}}{(J - \bar{j})[\text{H}]} \tag{6.2}$$

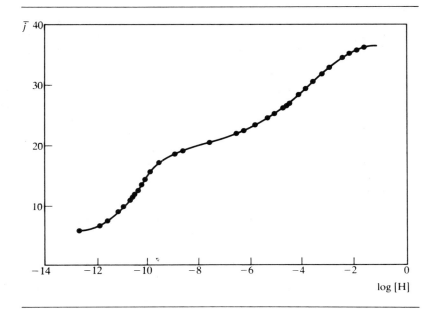

Fig. 6.1 Formation curve for the protonation of ribonuclease. (After C. Tanford, *The Physical Chemistry of Macromolecules*, Wiley, New York (1961).)

and, if K were indeed a constant, its value could be obtained from the intercept of the linear plot of $\log \bar{j}/(J - \bar{j})$ against pH. The term $\bar{j}$ has the same meaning as previously and may be obtained in the same way: by determining the free hydrogen ion concentration potentiometrically and using it to calculate the quantity $(H_t - [H])/A_t$. The assumption of a single equilibrium constant leads to the expression

$$\bar{j} = \frac{JK[H]}{1 + K[H]} \tag{6.3}$$

A plot of $\bar{j}J^{-1}$ against $\log [H]$ would therefore be of identical shape to the formation curve of a monobasic acid (see Figs 3.3 and 13.1) and the value of the single parameter K could be obtained by the curve-fitting procedure outlined in Chapter 3.

It is not surprising that this ultra-simple behaviour is never observed, even for those synthetic polymers, such as polymethacrylic acid, which contain identical function groups. Since protons are charged, protonation increases the positive charge on the macromolecule, and so discourages the addition of further protons. It would therefore be expected that an increase in $\bar{j}$ would lead to a decrease in K. The quantitative relationship between K and $\bar{j}$ would depend on the overall charge on the macromolecule at each value of $\bar{j}$; and this in turn is determined by the number of bound metal ions as well as by the degree of protonation. The variation of K with $\bar{j}$ also depends on the way in which the net charge is distributed, and so is affected by both the size and the shape of the molecule. But since like charges repel each other, the size also depends on the net charge. Macromolecules are known to expand, sometimes dramatically, as they become more ionised. Not even the shape remains constant. Changes in pH may result in the making, or breaking, of hydrogen bonds and so produce conformational changes. A macromolecule never, of course, has identical sites unless $\bar{j}$ equals either zero or J. Even if it contains only one type of functional group, the sites become different from each other as soon as any protonation has occurred; the distribution of charge at a free site which is next to a bound one may be very different from the distribution of charge at a free site which is surrounded by other free sites. Moreover, any change in the distribution of charge on the polymer may alter the activity coefficients of the ions in the adjacent solution.

The previous paragraph serves to outline the types of complication which occur in solutions of macromolecules with only one type of functional group. Most systems are much less simple than this. The fact that ribonuclease, for example, has at least seven types of site which can bind protons adds considerably to the already severe difficulties in interpreting values of $\bar{j}$, [H]. So the picture which emerges of the protonation of macromolecules involves a central species which may contain over one thousand identical sites, or a much smaller number of different sites (including, perhaps, two unique end-groups). As the degree of proton binding decreases, the charge on the macromolecules increases, but this is offset to some extent by the binding of other cations. With increasing net charge, the macromolecule expands, and maybe also undergoes conformational changes which may or may not be reversible. For the rest of the chapter, we shall consider briefly how far our original assumption of identical, independent binding sites can be modified to describe the complexities of reality.

6.2   The Coulombic correction

The effect of changing charge on macromolecule has been treated by dividing the free energy change of binding into two parts: a 'chemical' term, which represents the 'intrinsic' affinity between site and ion, and an 'electrostatic' term, which accounts for the work involved in moving a charged particle in an electrostatic field. Such division, although thermodynamically indefensible, is very convenient, particularly as the electrostatic contribution to the free energy change is often found to be proportional to the average net charge, $\bar{z}$. We may therefore write

$$\log K = \log K_i + C\bar{z} \tag{6.4}$$

where K_i is the 'intrinsic' protonation constant and the Coulombic constant C is the electrostatic proportionality factor, which depends on the temperature, ionic strength and the shape of the macromolecule. Following Tanford most authors write the term $C\bar{z}$ in equation (6.4) as

$$-2(\log e)w\bar{z} = -(\log e)(kT)^{-1}(\partial W_e/\partial \bar{z})_x$$

where k is the Boltzmann constant and x the dielectric constant. The term W_e is the electrostatic free energy of the polymer, and w is a parameter which depends on the electrostatic model used but is independent of the nature of the combining site.

The value of $\bar{z}$ may be obtained from $\bar{j}$, provided that the macromolecule binds only protons and that the value $\bar{j}_u$ of $\bar{j}$ for the uncharged form is known. In general, $\bar{z} = (\bar{j} - \bar{j}_u)$, and so $\bar{z}$ is a linear function both of $\bar{j}$ and of the ratio $\bar{j}/J$. For polyacrylic acids, the fully protonated form is uncharged and so $\bar{z} = (\bar{j} - J)$. for β-lactoglobulin, on the other hand, the neutral form contains about 54 bound protons, and can add another 40, so that $\bar{z}$ varies from -54 to $+40$ according to the relationship $\bar{z} = (\bar{j} - 54)$. If, as is likely, negatively charged macro ions bind metal ions as well as protons, the value of $\bar{z}$ can only be obtained if the total quantity of bound metal ion has also been determined, e.g. by equilibrium dialysis (see sect. 11.2).

Combination of equations (6.2) and (6.4) gives

$$\ln \frac{\bar{j}}{(J - \bar{j})[\mathrm{H}]} = \ln K_i + C\bar{z} \qquad (6.5)$$

so that, if the binding sites are indeed identical, a plot of the left-hand side of equation (6.5) against $\bar{z}$ will be a straight line of slope C and of intercept $\ln K_i$. If the linear plots of this type are obtained, we may infer both that the macromolecule has only one type of basic site and that its size and shape are

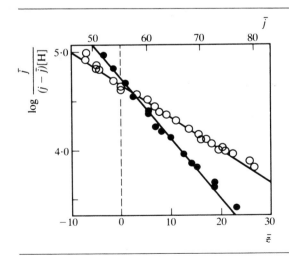

Fig. 6.2 Plot based on equation (6.5) for side-chain carboxyl groups of β-lactoglobulin. Full circles $I = 0.01$M; open circles $I = 0.15$M. (After Y. Nozaki, L. Bunville and C. Tanford, *J. Amer. Chem. Soc.*, **81**, 5527 (1959).)

little affected by protonation. Such behaviour is indeed observed for some rigid macromolecules, such as some globular proteins and polyacrylic acids (see Fig. 6.2). In these cases, it follows from equations (6.4) and (6.5) that

$$\bar{j} = \frac{JK_i\,e^{C\bar{z}}[H]}{1+K_i\,e^{C\bar{z}}[H]} \tag{6.6}$$

6.3   Further complications

There are several reasons for deviations from the behaviour described by equation (6.3).

1. The size of the macromolecule may vary with $\bar{j}$. Expansion often occurs as $\bar{z}$ increases and causes a corresponding change in the value of the Coulombic 'constant' C. If the macromolecule contains only one type of binding site, gradual expansion of a molecule causes smooth curvature in the plots of $\log \bar{j}/(J-\bar{j})[H]$ against $\bar{z}$ (see Fig. 6.3).

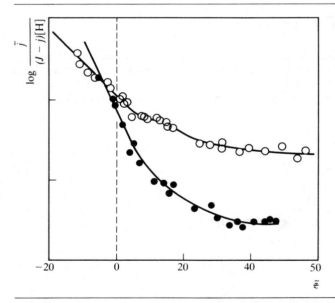

Fig. 6.3 Plot based on equation (6.5) for side-chain carboxyl groups of serum albumin. (After C.. Tanford, S. A. Swanson and W. S. Shore, *J. Amer. Chem. Soc.*, **77**, 6414 (1955).)

2. The shape of the macromolecule may vary with $\bar{j}$, on account of conformational changes, such as uncurling or refolding, which may again occur in species which contain only one type of functional group. Such changes have a more marked effect on protonic equilibria than those caused solely by expansion, partly because the 'electrostatic contribution' is no longer even roughly proportional to $\bar{z}$, and partly because the bound protons are likely to be involved in the conformational changes. If they do take part in these changes, the 'chemical contribution' to log K also varies with $\bar{j}$, and our original model of a single 'intrinsic' protonation constant for all binding sites can no longer be used, even if modified for Coulombic effects. In some macromolecules, conformational changes cause cooperative binding, with the result that K increases with $\bar{j}$. Thus the addition of a proton may be facilitated by the presence of a proton on a neighbouring site if, for example, two adjacently bound protons are necessary for the formation of a stabilising cross-linkage. Conformational changes cause marked changes of slope in the formation curve and in plots based on equation (6.5). If cooperative binding occurs, the plot of $\bar{j}J^{-1}$ against log [H] will be steeper than the formation curve for a monobasic acid.

3. Protons may be bound to more than one type of functional group. Indeed this is true for all naturally occurring acidic or basic macromolecules. As we have seen, ribonuclease has seven different types of groups which interact with protons. The average number of bound protons can then be represented by an expression analogous to (6.6) with one term for each type of protonation site, containing approximate values of J_A, K_{iA} and C_A for sites of type A, and so on. Thus

$$\bar{j} = \frac{J_A K_{iA} e^{C_A \bar{z}}[H]}{1 + K_{iA} e^{C_A \bar{z}}[H]} + \frac{J_B e^{C_B \bar{z}}[H]}{1 + K_{iB} e^{C_B \bar{z}}[H]} + \cdots \qquad (6.7)$$

and the formation curve shows a number of kinks (see Fig. 6.1). If only one type of group interacts with protons in a given range of pH, measurements of $\bar{j}$[H] in that region may be treated as if only one type of functional group were present. Additional complications arise, however, because the distribution of charge in the neighbourhood of a site depends not only on $\bar{z}$ and on the size and shape of the macromolecule, but also on the relative positions of the various types of functional group.

4. Other cations may compete with protons for basic sites. In principle, any site which can be protonated might also bind other cations. Purists could argue that the total acetate concentration $(Ac)_t$ is a buffer solution made up from sodium acetate and acetic acid should properly be written as

$$(Ac)_t = [Ac^-] + [HAc] + [NaAc] \qquad (6.8)$$

But since no evidence has been offered for the existence of undissociated NaAc in dilute aqueous solutions, the presence of the final term in the equation would seem to be not only pedantic but positively misleading. There may, indeed, be some slight interaction between Na^+ and Ac^- ions but the effect of this is relegated to the activity coefficient term and, properly, disappears at infinite dilution. Any difference between the value of pK for acetic acid in two different ionic media causes no surprise; it is accepted as a predictable 'acidity coefficient effect', which may well be partly caused by differences in interaction with counter-ions.

It is hardly surprising that protonic equilibria involving macromolecules which can carry widely varying numbers of charges should be particularly sensitive to the concentration of counter-ions. Not only does the ionic strength affect the activity coefficients of the macromolecule in a complex fashion, but binding of counter-ions reduces the number of sites available to protons, reduces the value of $\bar{z}$, alters the distribution of charge, and affects the ease of conformational changes. Competitive and ternary equilibria are discussed more fully in sect. 7.5 and 11.2; they are mentioned here merely to illustrate one of the many difficulties which beset those who attempt any quantitative study of equilibria involving macromolecules. Indeed, with all the complications outlined here, it is hardly surprising that the study of the protonation of macromolecules is a field in which the emergence of a numerical value of an equilibrium constant is something of a bonus. One is grateful if such difficult systems can be persuaded to yield up any information at all, whether it be a value of J, or an indication of changes in the type of protonation site, or of variations in size and shape.

Part III

Metal ion complexes

Chapter 7

Introduction to metal complex formation

As we saw in Chapter 1, equilibria in solution are by no means restricted to protonation reactions. Another large and very important group of equilibria are those which involve the formation of complexes between a metal ion and various Lewis bases known as ligands (i.e. 'binders').

7.1 Types of complex

A wide variety of anions and molecules can act as ligands, usually by donating an electron pair to the metal ion to form a σ bond. Although almost any metal ion, given a suitable ligand, can act as the central group of a complex, some form complexes more readily than others. Metal ions in certain parts of the periodic table have a marked preference for ligands with particular types of donor atom. Cations of the calcium group, for example, tend to form their most stable complexes with ligands which contain oxygen atoms, while ions of metals towards the end of the d-transition series form a wide variety of stable complexes with ligands which contain nitrogen atoms. Some ligands (such as those shown in Fig. 7.1) contain more than one donor atom, and so can be attached to the metal ion at more than one binding site.

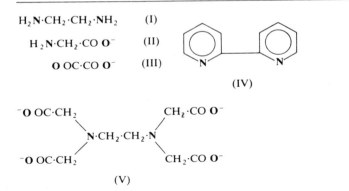

$$H_2N \cdot CH_2 \cdot CH_2 \cdot NH_2 \quad \text{(I)}$$

$$H_2N \cdot CH_2 \cdot CO \, O^- \quad \text{(II)}$$

$$O \, OC \cdot CO \, O^- \quad \text{(III)}$$

(IV)

$$^-O \, OC \cdot CH_2 \qquad\qquad CH_2 \cdot CO \, O^-$$
$$N \cdot CH_2 \cdot CH_2 \cdot N$$
$$^-O \, OC \cdot CH_2 \qquad\qquad CH_2 \cdot CO \, O^-$$

(V)

Fig. 7.1 Some chelating agents, with the potential donor atoms represented by boldface symbols. I. ethylenediamine ('en'); II. glycinate ('gly'); III. oxalate ('ox'); IV. 1.10-bipyridyl ('bipy'); V. ethylenediamine tetra-acetate ('EDTA').

They act as *chelating* agents, those with two donor atoms clipping on to the metal ion like the two pincers of a crab's claw, and those with three or more donor atoms acting like the multijaw grab on a crane. The number of ligands which can bind a central metal ion naturally depends on the number of donor atoms which are used by each ligand group, and by the preferred stereochemistry and relative sizes of both ligand and metal ion. Some typical complex ions are the following:

$$Fe(bipy)_3^{2+} \quad FeF_6^{3-} \quad Fe(CN)_6^{3-} \quad CuCl_4^{2-} \quad Ca(EDTA)^{2-} \quad Ni(en)_3^{2+} \quad Ag(NH_3)_2^+$$

Complexes need not, of course, be charged. Nor need they contain the maximum number of ligands which are capable of grouping around the particular metal ion. For example, if chloride ions are added to a solution of copper(II) nitrate, any of the species $CuCl^+$, $CuCl_2$, $CuCl_3^-$ and $CuCl_4^{2-}$ may be formed. The way in which the total copper(II) is distributed between the uncomplexed metal ion and the four possible chloro complexes depends on the concentration of free chloride ions and on the equilibrium constants for the formation of the complexes, in exactly the same way as the relative concentrations of the species PO_4^{3-}, HPO_4^{2-}, $H_2PO_4^-$ and H_3PO_4 in a mixture of phosphoric acid and strong base varies with the pH in a manner which depends on the values of the various protonation constants.

The addition of n ligands L to a central metal ion M is formally identical to the binding of j protons to the base A, the distribution of M between the various complexes ML_n or of A between the acids H_jA being dependent on the equilibrium concentration of either L or H. The algebra which is used in studies of metal complex equilibria therefore looks much the same as that which we have been using in our discussions of protonation equilibria. Chemically, of course, there are important differences between the two types of system. In protonation equilibria, the central group A acts as a base, donating one pair of electrons to each bound proton. In mononuclear metal complexes, however, the role of base (i.e. of Lewis base, or electron donor) is played by the ligand. The situation is different for the combination of metal ions with a protein. Since one protein molecule can bind a large number of metal ions, these now act as ligands and the central molecule of protein acts as a Lewis base, just as it does when it combines with protons.

A few species, such as the metal complexes $Fe_2(OH)_2^{4+}$ and $Ag_2Cl_n^{(n-2)}$ and the acetic acid dimer H_2Ac_2, contain two or more of both donor and acceptor groups. For such species, the labels 'ligand' and 'central group' are less useful, and the more general treatment outlined in Chapter 11 must be used.

As we saw in Chapter 1, the term 'complex' is itself misleading. Most equilibria involving metal ions are studied in aqueous solution, in which the metal ions are already strongly complexed – by water molecules. So-called complex formation occurs only if the added ligand binds to the metal ion even more strongly than water does, and so can displace some, or all, of the water from the hydration sphere of the metal ion. The number of water molecules displaced by one ligand group depends, of course, on the type and size of the oncoming ligand and on the number of donor groups it employs.

Provided that the measurements are made in dilute aqueous solutions at constant pressure, the concentration of water remains constant. The effect of the water in competing for binding sites on the metal ion therefore remains the same throughout a set of measurements; but it should nonetheless not be forgotten.

7.2   Stability constants

In this section, we shall assume that a 'complex' consists of a single metal ion, bound by at least one ligand L in addition to molecules of water. This simple treatment will exclude:

(a) polynuclear complexes, such as $Fe_2(OH)_2^{4+}$ which contain two or more metal ions; and

(b) mixed complexes, which contain more than one type of ligand in addition to water.

This latter group includes complexes, such as HgClBr and $FeCl(SCN)^+$, which are formed from two different added ligands L and L' and species such as $FeCl(OH)^+$ and $HFeCl_4$ in which the second 'complexing' ligand is derived from either H_2O or L by the addition or removal of a proton. Since detailed treatment of polynuclear, mixed and mixed-polynuclear complexes would upset the equilibrium of the book these species will be discussed only briefly (see Ch. 11). The basic algebra given below can, however, be used for metal ions which are hydrolysed, or polymerised, provided that they exhibit the same degree of hydrolysis, or polymerisation, over the whole range of conditions used. Thus the simplest forms of mercury(I), vanadium(IV) or uranium(VI) which exist in aqueous solution are the hydrated cations Hg_2^{2+}, VO^{2+} and UO_2^{2+}. Since these behave as single entities, and never dissociate, they may be treated as if they were simple metal ions.

Unless it is stated otherwise, we shall optimistically assume that activity coefficients can be kept constant throughout a complete series of measurements. Experiment, as well as common sense, suggests that this optimism is most likely to be justified if the composition of the medium is varied as little as possible; and if such changes as are made involve species which are not too dissimilar from each other. A constant ionic medium should always be used for studies of metal complex formation if at all possible because it is even more difficult to disentangle the effect of variation of activity coefficients on these reactions than on protonation equilibria. There are several reasons for this. Since many metal ions can coordinate four or six donor atoms, species in which the central group is bound to only one or two ligands are much rarer among metal complexes than among acids. Most complexes formed by metal ions therefore fall into the 'several ligands' category. Interpretation of measurements may be made difficult by the fact that successive equilibrium constants for metal complex formation often lie more closely together than do adjacent values of protonation constants, so that data cannot be divided into sections which each refer to the formation of a single complex. As we shall see (sect. 7.5), most species which act as Lewis bases and complex with metal ions also act as Brönsted bases and combine with hydrogen ions. Complex formation between metal and ligand then involves competition between metal ions and protons for the basic species L. This situation can be exploited to allow us to measure the equilibrium concentration of L, but it has the disadvantage of increasing the number of different species which are present in solution. If activity coefficients are not controlled, they will vary not only with the total ionic strength, but also with the size, charge and chemical nature of each ion concerned. So the greater the diversity of the solute species, the stronger is the case for trying to keep the activity coefficients constant. Ways of approaching this unattainable state are discussed in sect. 7.5. Here we shall merely assume that 'equilibrium constants' which are expressed as concentration quotients are indeed constants.

We have thus restricted ourselves to discussing reactions of the type

$$M \text{ aq} + n L \text{ aq} = ML_n \text{ aq} + z_n \ H_2O \tag{7.1}$$

for which the equilibrium constant

$$\beta_n = \frac{[ML_n \text{ aq}][H_2O]^{z_n}}{[M \text{ aq}][L \text{ aq}]^n} \tag{7.2}$$

is the overall stoichiometric stability constant (or formation constant) of the complex ML_n aq. As before, charges are omitted. As an alternative to equation (7.1), we may consider the reaction

$$ML_{n-1} \text{ aq} + L \text{ aq} = ML_n \text{ aq} + (z_n - z_{n-1})H_2O \tag{7.3}$$

in which the species ML_n aq is formed by the addition of one ligand to the preceding complex ML_{n-1} aq. The corresponding equilibrium constant

$$K_n = \frac{[ML_n \text{ aq}][H_2O]^{(z_n - z_{n-1})}}{[ML_{n-1} \text{ aq}][L \text{ aq}]} \tag{7.4}$$

is the stoichiometric stepwise stability (or formation) constant of ML_n. It follows that

$$K_n = \beta_n / \beta_{n-1} \tag{7.5}$$

and, conversely, that

$$\beta_n = K_1 K_2 K_3 \ldots = \prod_1^n K_n \tag{7.6}$$

It is usually, although tacitly, assumed that the water is at a constant concentration, which is, moreover, its standard state. The first assumption is surely justified for any series of solutions in which ionic activity coefficients are constant; and, if the first requirement is achieved, the second is merely a question of definition. All reference to water in the expressions (7.1) to (7.4) can then be dropped, and the overall and stepwise stability constants defined much more elegantly as

$$\beta_n = [ML_n]/[M][L]^n \tag{7.7}$$

and

$$K_n = [ML_n]/[ML_{n-1}][L] \qquad (7.8)$$

The role which water plays must not, however, be forgotten when it comes to interpreting values of stability constants, particularly if we wish to compare stabilities of complexes formed by two chemically similar ligands which form complexes containing different numbers of chelate rings (and therefore different numbers of bound water molecules), or series of measurements made at different pressures, or in partly-aqueous mixed solvents.

There will also be interaction between the various species we are studying and the components of the bulk electrolyte. Any observed stability constant therefore depends on the free energy change which accompanies the transfer of a metal ion M from an environment of solvent molecules and bulk electrolyte ions to one in which it is in close proximity to at least one ligand L. So a stability constant is, rigorously, an equilibrium constant for the displacement, by ligand, of an unknown number of solvent molecules and background ions from the metal ion M.

When M interacts strongly with L, measurable complex formation occurs at such low values of [L] that composition of the solution need not differ grossly from that of the pure ionic medium. An effectively constant concentration of background electrolyte allows us to measure complex formation between M and L relative to a constant level of interaction between M and the bulk electrolyte; or, to put it another way, the use of a constant ionic medium enables us to hold activity coefficients constant.

The study of weak complexes is often more difficult because higher values of [L] are needed and we may no longer be justified in assuming that activity coefficients remain constant. The observed effect of an increase in [L] may then be due either to increased complex formation between M and L, or to variation in activity coefficients or to a combination of these factors; and there is no rigorous way in which an observed change may be apportioned between these two effects. If a complex is extremely weak, its very existence is a question of how we define 'close proximity', and the value of its stability constant depends on the distance we assign to the separation between M and L in the complex. In 'outer sphere' complexes, in which the Coulombic interaction occurs between an anionic ligand and a metal ion which retains its primary solvation sphere, stability constants have been calculated assuming interionic distances as large as 14 Å.

7.3 Inert complexes

For a small minority of metal ions, of which the most notable are Cr^{3+} and Co^{3+}, equilibrium between the aquo metal ion, the ligand and the complex is established extremely slowly, regardless of the (thermodynamic) stability of the complex. Once equilibrium has been reached, it may then be possible to use traditional quantitative methods to determine the resultant concentrations of some or all of the species which have been formed. Addition of an analytical reagent does, of course, remove one of the species by precipitation or complex formation; but the equilibrium readjusts so slowly that the concentrations of all the other species do not change appreciably within the time taken to make the necessary measurements. A beautiful study of chromium(III) thiocyanate complexes was carried out in this way as early as 1914 by Niels Bjerrum, who analysed the equilibrium concentrations of each of the eight species SCN^-, Cr^{3+} and $Cr(SCN)_n^{(3-n)+}$ (where $n = 1, 2, 3, 4, 5$ and 6). Complexes of the few other metal ions (e.g. Co(III), Pt(II) and Pt(IV)) which undergo slow substitution of the aquo complex may be studied in a similar way. But most solutions containing metal complexes, like acid–base systems, come to equilibrium extremely quickly. Of the great majority of complexes which are formed within the time taken to mix two solutions, most have average lifetimes of less than 10^{-2} s and some are as short-lived as 10^{-9} s. It is therefore seldom possible to add analytical reagents, and metal complex formation must normally be monitored by measuring physical quantities which respond to the equilibrium concentration of one of the species, or to some property of the solution as a whole, without displacing the equilibrium. Methods which give the concentration of a single species are much to be preferred, particularly if several complexes coexist; and it is these methods which are outlined below.

By no means all inert systems may be studied easily; some are extremely resistant to investigation. There is a legend that, at the outbreak of the First World War, Niels Bjerrum was attempting to study the hydrolysis of chromium(III), but by Armistice Day, the solutions had still not come to equilibrium. A particularly recalcitrant system is iron(III) cyanide, because the complexes are very stable, as well as being very inert. Addition of acid, with a view to establishing competitive equilibria with protons (cf. sect. 7.5), causes the slow escape of gaseous HCN. So it is not surprising that the stepwise stability constants of this system are as yet unmeasured. The value of β_6 has, however, been obtained indirectly, via Hess's law, from the calorimetric value of the heat of formation of the hexocyano complex in aqueous solution, together with the entropies of the aqueous metal ion, ligand and complex.

7.4   Basic algebra

The algebraic procedures for handling stepwise equilibria in solutions containing metal complexes are, not surprisingly, extremely similar to those we have used handling other types of multiple equilibria. We shall assume for the moment that we are dealing with one of the few ligands, e.g. the chloride ion, which does not combine with protons. This allows us to follow the drill which is familiar from Chapter 5, viz:

   (i) to express the total concentrations of both central group and ligand (here, metal ion and chloride ion) in terms of the equilibrium concentrations of the individual species present;

(ii) to combine these two total concentrations with the concentration of free ligand in order to obtain the average number of ligands bound to each metal ion;

(iii) to replace the values of the various equilibrium concentrations by the appropriate expression involving the equilibrium concentrations only of the ligand and metal ion, together with the stability constant of the particular complex.

We then find that, knowing only the total concentrations of metal ion and ligand, and the free concentration of ligand, we have obtained an expression from which the stability constants may be obtained.

The total concentrations

$$M_t = [M] + [ML] + [ML_2] + \cdots + [ML_N] \tag{7.9}$$

$$= \sum_0^N [ML_N] \tag{7.10}$$

$$= \sum_0^N \beta_n [M][L]^n \tag{7.11}$$

and

$$L_t = [L] + [ML] + 2[ML_2] + \cdots + N[ML_N] \tag{7.12}$$

$$= [L] + \sum_0^N n[ML_n] \tag{7.13}$$

$$= [L] + \sum_0^N n\beta_n [M][L]^n \tag{7.14}$$

are known from the way in which the solutions have been prepared. The average number of ligands bound to each metal ion is, of course, the average value of n, and is given by

$$\bar{n} = \frac{\text{concentration of ligand bound to metal}}{\text{total concentration of metal}} \tag{7.15}$$

$$= \frac{L_t - [L]}{M_t} \tag{7.16}$$

$$= \frac{[ML] + 2[ML_2] + \cdots + N[ML_N]}{[M] + [ML] + [ML_2] + \cdots + [ML_N]} \tag{7.17}$$

$$= \frac{\beta_1[M][L] + 2\beta_2[M][L]^2 + \cdots + N\beta_N[M][L]^N}{[M] + \beta_1[M][L] + \beta_2[M][L]^2 + \cdots + \beta_N[M][L]^N} \tag{7.18}$$

For such systems in which no complex contains more than one metal ion, the free concentration of metal ion appears (raised only to the first power) in each term, and so may be cancelled out to give the relationship

$$\bar{n} = \frac{\beta_1[L] + 2\beta_2[L]^2 + \cdots + N\beta_N[L]^N}{1 + \beta_1[L] + \beta_2[L]^2 + \cdots + \beta_N[L]^N} \tag{7.19}$$

$$= \frac{\sum\limits_{0}^{N} n\beta_n[L]^n}{\sum\limits_{0}^{N} \beta_n[L]^n} \tag{7.20}$$

which is a polynomial in the single variable [L]. The values of the stability constants β_n may therefore be obtained by solving sets of equations obtained by substituting pairs of experimental values of $\bar{n}$ and [L] into equation (7.20). Since the values of $\bar{n}$ are obtained from equation (7.16), and since values of M_t and L_t are usually known, we need to measure only one equilibrium concentration, [L]. Methods of solving equation (7.20) in order to obtain values of β_n from the function $\bar{n}([L])$ are the same as those used for obtaining values of protonation constants from the exactly analogous function $\bar{j}([H])$ (see Ch. 5).

Although $\bar{j}$ is the most important secondary concentration variable used in studies of protonation equilibria, we also mentioned the variables α_c, defined as that fraction of the central group which exists in the form of a particular acid H_jA. Similar secondary concentration variables α_c may be defined for metal–ligand equilibria such that α_c is that fraction of the total metal ion which is complexed by c ligands. The variable α_c, like $\bar{n}$, may be expressed in terms both of measurable quantities and of the stability constants which we might wish to evaluate. We shall suppose that we have some experimental technique which gives us the equilibrium concentration of the complex ML_c, and that, as before, the value of M_t is known. The value of

$$\alpha_c = \frac{[ML_c]}{M_t} \tag{7.21}$$

may then be calculated. Use of the expressions (7.7) for $[ML_c]$ and (7.11) for M_t gives

$$\alpha_c = \frac{\beta_c[M][L]^c}{[M] + \beta_1[M][L] + \beta_2[M][L]^2 + \cdots + \beta_N[M][L]^N} \tag{7.22}$$

For systems which contain only one metal ion per complex, we may again divide top and bottom of the fraction by [M] to give

$$\alpha_c = \frac{\beta_c[L]^c}{1 + \beta_1[L] + \beta_2[L]^2 + \cdots + \beta_N[L]^N} \tag{7.23}$$

$$= \frac{\beta_c[L]^c}{\sum\limits_{0}^{N} \beta_n[L]^n} \tag{7.24}$$

The function $\alpha_c([L])$, like $\bar{n}([L])$, is a polynomial in the single variable $[L]$ and so can be solved to give values of the stability constants β_n. It differs slightly from $\bar{n}([L])$, at least in principle, in that it is based on the measurement of the equilibrium concentrations of two species, ML_c and L, whereas $\bar{n}([L])$ requires values only of $[L]$. In practice, however, the function α_c is often used for systems in which the total concentration of metal ion is very much less than that of ligand, so that the value of $[L]$ approximates to the total concentration, L_t, of ligand. The function $\alpha_c([L])$, like $\bar{n}([L])$, can then be calculated from known values of M_t and L_t together with the measured values of the equilibrium concentration of only one species. We shall see that the quantity α_0 (which represents that fraction of metal which is uncomplexed) is particularly useful since it can be measured for many metal ions with a metal-, amalgam- or redox-electrode (Ch. 8). When $c = 0$, equation (7.24) reduces to

$$\alpha_0 = \frac{1}{\sum\limits_0^N \beta_n [L]^n} \tag{7.25}$$

The value of α_c for an uncharged complex may sometimes be determined by liquid–liquid partition or, indirectly, determination of solubility (see Ch. 9). When only one complex is formed, its concentration, and hence the appropriate value of α_c, may sometimes be measured spectrophotometrically (see Ch. 10).

Since the value of α_c depends on the parameters β_n and on the single variable $[L]$, the stability constants may be obtained by substituting pairs of values of α_c and $[L]$ into equation (7.24) and solving the simultaneous equations so obtained. The principle is exactly the same as the calculation of stability constants from data $\bar{n}$, $[L]$. Methods of processing the functions $\bar{n}([L])$ and $\alpha_c([L])$ are discussed in more detail in Chapter 12.

7.5 Real-life algebra: competition with protons

A glance through the compilations of stability constants of metal complexes soon reveals that very few values have been obtained from direct measurements of the concentration of a complex or of the free ligand, or even of the uncomplexed metal ion. Overwhelmingly the most popular method is the measurement of hydrogen ion activity or concentration by means of the glass electrode. Any probe for hydrogen ions can, of course, provide the experimental entry into the melée of competing equilibria which are set up in most real-life solutions containing metal complexes. Since the ligand L is usually a Brönsted base, it may become protonated in solution to form one or more of the acids H_jL in addition to displacing water molecules from a hydrated metal ion to form any, or several, of the complexes ML_n. The overall formation constant of an acid which contains j protons will be written as β_j^H, and the overall stability constant of a metal complex which contains n ligands as β_n, to allow us to distinguish between two constants, such as β_2 and β_2^H, which refer to the addition of the same number of groups in different systems. The possibility of any variation in activity coefficients or of the formation of mixed or polynuclear species will again be ignored.

Although the presence of two sets of complexes necessitates the use of heavier algebra, the basic procedure is exactly the same as that used in the previous section. First, the total concentrations of the metal ion, ligand and hydrogen ion are expressed in terms of the equilibrium concentrations of the species they contain. The value of M_t is again given by equation (7.11) but the total concentration of ligand now includes the various protonated species H_jL, so that

$$L_t = [L] + ([ML] + 2[ML_2] + \cdots + N[ML_N])$$
$$+ ([HL] + [H_2L] + \cdots + [H_JL]) \tag{7.26}$$

$$= [L] + \sum_1^N n[ML_n] + \sum_1^J [H_jL] \tag{7.27}$$

The total concentration of dissociable hydrogen ions is given by equation (5.2) as

$$H_t = [H] - [OH] + ([HL] + 2[H_2L] + \cdots + J[H_JL]) \tag{7.28}$$

$$= [H] - K_w[H]^{-1} + \sum_0^J j[H_jL] \tag{7.29}$$

The average number of ligands bound to each metal ion is again given by equation (7.15), where the concentration of 'ligand bound to metal' is now given by

$$\sum_0^N n[ML_n] = L_t - \sum_0^J [H_jL] \tag{7.30}$$

From equation (5.7) we have

$$\bar{j} \sum_0^J [H_jL] = \sum_0^J j[H_jL] \tag{7.31}$$

Combination of equations (7.15), (7.28), (7.30) and (7.31) gives

$$\bar{n} = \frac{L_t - \bar{j}^{-1}(H_t + [H] - K_w[H]^{-1})}{M_t} \tag{7.32}$$

This expression (7.32) contains:

(i) the three total concentrations of M, L and H, which are usually known;
(ii) one equilibrium concentration [H], which can readily be measured;
(iii) the secondary concentration variable, $\bar{j}$, which may be calculated from the value of [H] provided that the protonation constants β_i^H have been determined under the same experimental conditions.

Measurement of [H] therefore gives values of $\bar{n}$; but in order to calculate stability constants from values of $\bar{n}$, the corresponding values of the free ligand concentration must be known. Now the fraction of uncomplexed L which is also unprotonated is given by

$$\alpha_0^H = \frac{[L]}{\sum\limits_{0}^{J}[H_jL]} = \frac{[L]}{L_t - \bar{n}M_t} \tag{7.33}$$

The value of [L] may then be calculated as

$$[L] = \alpha_0^H (L_t - \bar{n}M_t) \tag{7.34}$$

where the required value of α_0^H (like that of $\bar{j}$) can be obtained from the measured value of [H] and known values of the protonation constants (see equations (4.24) to (4.26) and compare equation (7.24)). Measurements of the single variable [H] then give pairs of values of both $\bar{n}$ and [L] provided that the total analytical concentrations and the protonation constants of the ligand are known.

7.6   Preparing the solutions

It is obvious that measurements should be of the highest possible precision if it is hoped to obtain reliable values of stability constants from solutions in which a variety of species coexist. And good measurements depend on good starting materials as well as on good measuring equipment and technique. The initial solutions should be as free as possible from impurities which might interfere either with the analysis or with the equilibria which are being studied; and the total concentrations H_t, L_t and M_t must be determined with both accuracy and precision. But before solutions can be prepared or analysed, they must of course be designed. A solution of 'metal ions' must contain the equivalent number of counter-ions. What anions should be used? If activity coefficients are to be held as constant as possible, what background salt should be added. And at what concentration?

The counter-ion should be chosen so that it upsets the measurements as little as possible; the main factors to be considered are the extent to which it associates with the metal ion and the solubility of any compounds it may form with species which are present in solution. We have seen that it is naïve to think of complex formation between M and L as the simple addition reactions

$$M + nL \rightleftharpoons ML_n$$

rather than as the replacement reactions

$$M.mH_2O + nL.lH_2O \rightleftharpoons ML_n.cH_2O + (m + nl - c)H_2O$$

But we approach reality even more closely if we envisage the ligand L as competing for a site on the metal ion with either a water molecule, or a counter-ion, or both. The complexes ML_n, as well as the free metal ion, may be associated with both counter-ions and solvent molecules, though since any Coulombic attraction decreases with decreasing charge, interaction between complexes and counter-ions will be more important for uncharged ligands than for anionic ones. If the object of studying equilibria in solution is to gain information about interaction between M and L, the competition between L and counter-ions is merely a nuisance; but we can minimise its effect by choosing a counter-ion which has the lowest possible tendency to associate with metal ions. If the interaction is primarily electrostatic, it would be weakest for singly charged anions with the largest possible radius. The perchlorate ion (ClO_4^-, of radius ~ 0.235 nm) is often used. Sometimes, however, the perchlorates of bulky complex cations such as $Ni(en)_3^{2+}$ are only sparingly soluble especially in mixed aqueous–organic solvents. It may then be more convenient to use the nitrate ion (NO_3^- of radius ~ 0.189 nm) as counter-ion.

It is obviously sensible to use the same species both as counter-ion and as the anion of background electrolyte. There remains the choice of the bulk cation, which, like the anion, should have as low a tendency as possible to form complexes. Sodium is a popular bulk cation, as it forms very weak complexes and its perchlorate is extremely soluble in water and in partly aqueous solvents. However, lithium perchlorate, although much less soluble in water, has advantages in studies of equilibria which involve protons, on account of the similarity between the H^+ and Li^+ ions.

The concentration of background electrolyte depends on the use to which the measurements are being put. The chemist's top priority is often that activity coefficients should be kept as nearly constant as possible; and this is most readily achieved by the use of a very high concentration of bulk electrolyte. The variation of M_t, H_t and L_t during the measurements then causes little change in the overall nature of the solutions, and there is some hope that all reacting species behave as if they were at infinite dilution in that particular ionic medium. A 3M solution of sodium perchlorate is often used for this type of work. Biochemists, on the other hand, often wish to study equilibria under conditions which are comparable with those which take place in living cells, in which the ionic strength is of the order of 0·1M. They prefer to risk greater (though, they hope, not excessive) variation in activity coefficients in order to obtain results which are optimally applicable to the systems that most interest them.

The concentration of bulk electrolyte can be held 'constant' in a number of slightly different ways. Let us suppose that the background electrolyte is 3M sodium perchlorate and that we wish to study complex formation between a metal ion M^{z+} and an anionic ligand L^{y-}, by adding $M(ClO_4)_z$ to Na_yL. There is then the choice of designing the solutions so that, for example, $[Na^+] = 3M$, or $[ClO_4^-] = 3M$, or that ($[Na^+] \nleq 3M$ and $[ClO_4^-] \nleq 3M$). Table 7.1 gives details of the concentrations of metal ion solution $[M^{z+}] = M_t$ and ligand solution $[L^{y-}] = L_t$ in the three cases mentioned, and shows how the concentrations change when the solutions are mixed.

Table 7.1 Molarities of some differently prepared ionic media.

	Solution 1. Metal ion			Solution 2. Ligand			Mixture of v_1 ml of solution 1 and v_2 ml of solution 2			
	M_t	$[Na^+]$	$[ClO_4^-]$	L_t	$[Na^+]$	$[ClO_4^-]$	M_t	L_t	$[Na^+]$	$[ClO_4^-]$
$[Na^+]=3$	M_i	3	$3+zM_i$	L_i	3	$3-yL_i$	$\dfrac{v_1 M_i}{v_1+v_2}$	$\dfrac{v_2 L_i}{v_1+v_2}$	3	$3+\dfrac{zM_i v_1+yL_i v_2}{v_1+v_2}$
$[ClO_4^-]=3$	M_i	$3-zM_i$	3	L_i	$3+yL_i$	3	$\dfrac{v_1 M_i}{v_1+v_2}$	$\dfrac{v_2 L_i}{v_1+v_2}$	$\dfrac{yL_i v_2-zM_i v_1}{v_1+v_2}$	3
$[Na^+]\not<3$ $[ClO_4^-]\not<3$	M_i	3	$3+zM_i$	L_i	$3+yL_i$	3	$\dfrac{v_1 M_i}{v_1+v_2}$	$\dfrac{v_2 L_i}{v_1+v_2}$	$3+\dfrac{yL_i v_2}{v_1+v_2}$	$3+\dfrac{zM_i v_1}{v_1+v_2}$

The materials used should be of the highest possible purity; time and money expended on preparing good solutions are well invested. Solutions of metal perchlorates may readily be prepared by dissolving the metal carbonate or oxide in a small excess of reagent-grade perchloric acid, and subsequently analysed for metal ion and acid. Attention should, of course, also be paid to the purity of the solvent. Water and alcohols should be doubly distilled from a hard-glass vessel, and dioxan purified by fractional crystallisation.

Solutions for analysis and measurement are normally prepared by dilution of standard stock solutions. All solutions should be stored at the constant temperature at which measurements will be made.

For work at 'room' temperature the whole laboratory can often be thermostatically controlled at, say, $25° \pm 1°C$. The temperature of the thermostat tank in which the measurements are done can then usually be kept steady to $\pm 0.1°C$.

The stock solutions (including those prepared by weighing-out of reagent-grade materials) should always be analysed, preferably by two widely differing methods such as volumetric and gravimetric techniques. The analysis of stock solutions is often considered to be one of the most tedious aspects of the determination of stability constants. But since the values of M_t, H_t and L_t form the basis of all subsequent calculations, the utmost care should be taken with the initial analyses. Failure to obtain 'good' stability constants is often traceable, after much elapsed time, to inadequate analysis of stock solutions.

Dilute solutions should be prepared by using Grade A volumetric glassware. (Some workers may feel that the effort involved in recalibration of Grade A glassware is amply repaid by their resulting sense of virtue; but it is questionable whether it leads to any significant improvement in the values of the stability constants.) In order to allow for any temperature change which may occur during dilution, slightly less than the required volume of diluent should be added in the first instance, and the dilution completed by topping up to the mark only when the bulk of the diluted solution has returned to the constant temperature of the surroundings.

In addition to conventional methods of analysis, there are a number of useful tricks of the equilibrium chemist's trade.

The stock solution of strong acid (often $HClO_4$) may conveniently be analysed potentiometrically, using Gran's method. A volume V of acid of initial concentration H_i is titrated with a volume v of strong base, of concentration B_i, in a cell which consists of an electrode which is reversible to hydrogen ions in combination with a reference electrode. In the acidic and alkaline regions, respectively, we may define two functions, ϕ and ϕ', such that

$$\phi = (V + v)10^{EF/2 \cdot 303RT} \tag{7.35}$$

and

$$\phi' = (V + v)10^{-EF/2 \cdot 303RT} \tag{7.36}$$

where E is the e.m.f. of the cell and F is the faraday. Since the quantity $10^{EF/2.303RT}$ is directly proportional to the free hydrogen ion concentration (see Ch. 8), we may write

$$\phi \propto (V+v)[H] \tag{7.37}$$

$$\phi' \propto (V+v)[OH] \tag{7.38}$$

In acidic solutions, before the end-point has been reached, we have

$$[H] = (H_i V - B_i v)/(V+v) \tag{7.39}$$

and, when excess of alkali has been added

$$[OH] = (B_i v - H_i V)/(V+v) \tag{7.40}$$

Thus the quantities ϕ and ϕ' are both linear functions of v, and both become zero when v is the titre at the equivalence point. The plots $\phi(v)$ and $\phi'(v)$ should therefore intersect each other, and the x-axis, at the equivalence point (see Fig. 7.2). (Intersection below the x-axis indicates that the alkali is contaminated with carbonate. The plot $\phi'(v)$ must then be ignored and the equivalence point obtained from the point of intersection of the line $\phi(v)$ with the x-axis.) Gran's method is quick, convenient and accurate, and can also be adapted to other systems, such as weak acids, strong acids in the presence of a hydrolysable ion, and metal ion redox titrations.

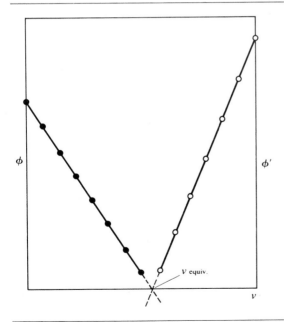

Fig. 7.2 Gran plot (cf. equations (7.35) and (7.36)) for titration of strong acid with v ml of strong base. Full circles, ϕ; open circles, ϕ'.

Cation exchange resins in the hydrogen form are useful for analysis of stock solutions of bulk electrolyte, and for mixtures of metal and hydrogen ions. If a portion of solution contains H moles of protons and M moles of metal ion M^{z+}, the solution emerging from an H–cation exchange column will contain $(H + zM)$ moles of H. When the stock solution has been analysed for metal ion alone, the concentration of acid can be found.

Measurements which involve variation in H_t are often carried out by adding a solution of base. Addition of alkali, such as sodium hydroxide, is liable to cause local precipitation of metal hydroxide, which disperses very slowly, or not at all. This nuisance may be avoided by using a solution of sodium hydrogen carbonate, instead of the hydroxide, provided that measurement is restricted to fairly acidic solutions (pH $\leqslant 5$).

Occasionally, the analysis of a solution of an organic base, such as a carboxylate ion, can be carried out potentiometrically in the same operation as the measurement of $\bar{\jmath}$. A volume V of a solution of, say, NaL of initial concentration L_i is titrated with a volume v of strong acid of concentration H_i, and the free hydrogen ion concentration is measured potentiometrically. From equations (3.12) and (4.18), the value of $\bar{\jmath}$ in an acidic solution is given by

$$\bar{\jmath} = \frac{H_t - [\mathrm{H}]}{L_t} = \frac{H_i v - [\mathrm{H}](V + v)}{L_i V} \tag{7.41}$$

Since, for a monobasic acid, $\bar{\jmath}$ tends to a value of $\bar{\jmath}_{max} = 1$ at high acidities, the required value of L_i is obtained as

$$L_i = \lim_{[\mathrm{H}] \to \infty} \{H_i v - [\mathrm{H}](V + v)V^{-1}\} \tag{7.42}$$

Once we have prepared, and analysed, the stock solutions with the greatest possible care we can use them for studying metal complex formation by one of the techniques described in Chapters 8 to 11.

Chapter 8

Experimental methods: I. E.m.f.

The ideal method for studying equilibria in solution should give accurate and precise values of the free concentrations of all species present in any conceivable system without disturbing the position of equilibrium; and it should be quick, foolproof and fun to use. Not surprisingly, no such technique has been devised. Of the methods available, potentiometry is the most versatile and most precise. Occasionally, the e.m.f. of a cell can lead directly to values of [L] which, when combined with the total concentrations M_t and L_t, yield the function $\bar{n}([L])$. Potentiometric measurements of [M] give the function $\alpha_0(L_t)$ from which the function $\alpha_0([L])$ can be obtained by iteration. Equilibria in non-aqueous organic solvents, and in fused salts, as well as aqueous systems, have been studied in this way. But by far the greatest number of potentiometric studies of complex formation involve measurement of the pH of aqueous solutions. Combination of values of [H] or {H} with those of H_t, L_t, M_t and β_i^H lead to the function $\bar{n}([L])$.

8.1 Basic theory

Typical potentiometric cells, suitable for titrations, are shown in Fig. 8.1. The composition of the test solution can be varied by addition of titrants from burettes. Immersed in the test solution are two 'electrodes', more precisely described as: (i) the probe(s) for the species being monitored; and (ii) a liquid junction J leading to a reference half-cell via a salt bridge. We shall arbitrarily write the cell as

$$-\text{reference half-cell} \,|\, \text{salt bridge} \,|\, \text{test solution} \,|\, \text{probe} + \qquad (8.1)$$
$$\phantom{-\text{reference half-cell} \,|\,} \text{j} \text{J}$$

with the probe as the more positive electrode.

We shall again optimistically assume that activity coefficients can be kept constant and, by a suitable definition of standard state (see p. 36 and sect. 7.6), we shall set them equal to unity. The Nernst equation (cf. equation (8.9)) for electrode response can then be expressed in terms of concentrations rather than activities.

The e.m.f. of cell (8.1) is given by

$$E_{\text{cell}} = E_p - E_{\text{ref}} + E_J + E_j \qquad (8.2)$$

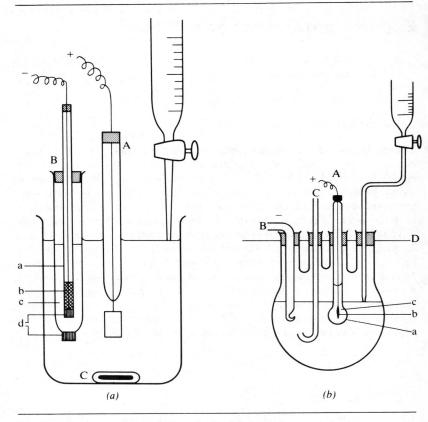

Fig. 8.1 Assemblies for potentiometric titration.
(*a*) Simple, open beaker cell.
   A. Platinum flag probe electrode (e.g. redox or quinhydrone).
   B. Calomel reference electrode.
      a. Pt wire; b. $Hg–Hg_2Cl_2$ paste; c. KCl solution; d. porous plugs.
   C. Magnetic stirrer.
(*b*) Cell for thermostatted titration under nitrogen.
   A. Glass membrane probe electrode.
      a. active membrane; b. internal Ag–AgCl reference electrode; c. aqueous HCl.
   B. salt bridge leading to reference half-cell.
   C. inlet for nitrogen.
   D. level of thermostat bath.
(After F. J. C. Rossotti and H. Rossotti, *The Determination of Stability Constants*, McGraw-Hill, New York (1961).)

where E_p and E_{ref} are respectively the potentials of the probe and reference half-cells, and E_J and E_j are the diffusion potentials at junctions J and j. As will be shown below (cf. equation (8.10) and Table 8.1), many probes respond to the concentration of a single species S according to the relationship

$$E_p = E_p^0 + \lambda_s RTF^{-1} \ln [S] \tag{8.3}$$

where E_p^0 and λ_s are constants. Since values of E_{ref} and E_j are unaffected by changes in the test solution, we may write the e.m.f. of cell (8.1) as

$$E_{cell} = E^0 + \lambda_s RTF^{-1} \ln [S] + E_J \tag{8.4}$$

where the value of

$$E^0 = E_p^0 + E_{ref} + E_j \tag{8.5}$$

is constant. It is often convenient to combine E^0 and E_J to give

$$E^{0'} = E^0 + E_J \tag{8.6}$$

which will also be constant provided that E_J does not vary. If this condition is fulfilled equation (8.4) becomes

$$E_{cell} = E^{0'} + \lambda_s RTF^{-1} \ln [S] \tag{8.7}$$

The value of [S] may then be readily calculated from the observed e.m.f. of cell (8.1). But before we assess whether or not we are over-optimistic in hoping that we can keep $E^{0'}$ constant we must consider how a potential arises at a liquid–liquid junction.

If two solutions are in contact, each solute species tends to diffuse away from that region where its activity is highest. The potential which arises at the boundary between two ionic solutions (or between an ionic solution and a non-ionic one) is due to the fact that different ions travel at different speeds. Suppose that a solution of an electrolyte A^+B^- is in contact with a more dilute solution of the same solute. Both types of ion will diffuse across the boundary into the more dilute solution. If the mobility of A^+ is higher than that of B^-, more cations than anions will pass into the more dilute solution in a given time, and so cause a small separation of charge at the boundary. This liquid junction potential contributes to the observed e.m.f. of the cell. Methods of estimating its value and of minimising the nuisance it causes are discussed in sect. 8.4.

8.2 Probes based on electron transfer

In this section we shall discuss electrodes which take part in an oxidation-reduction process. Probes based on membrane action are treated in sect. 8.3.

Electrode half-reactions may conveniently be written in the general form

$$\sum_I \lambda_I [I] + \sum_U \lambda_U [U] + \varepsilon \to 0 \tag{8.8}$$

where I and U represent ions and uncharged species respectively and the stoichiometric coefficients λ are positive for oxidised species and negative for

Table 8.1   Some probe electrodes.

Type of electrode	Half-reaction*	λ_I values	λ_U values	$F(E_p - E_p^{\ominus})/RT$	$E_p^{\ominus}$
Hydrogen gas (platinum black)	$H^+ - \tfrac{1}{2}H_2 + \varepsilon = 0$	$\lambda_{H^+} = 1$	$\lambda_{H_2} = -\tfrac{1}{2}$	$\ln[H^+]$	$-\tfrac{1}{2}RTF^{-1}\ln p_{H_2}$
Metal/metal ion	$Ag^+ - \boxed{Ag} + \varepsilon = 0$	$\lambda_{Ag^+} = 1$	$\lambda_{Ag} = -1$	$\ln[Ag^+]$	$E_p^{\ominus}$
	$\tfrac{1}{2}Zn^{2+} - \tfrac{1}{2}Zn(Hg) + \varepsilon = 0$	$\lambda_{Zn^{2+}} = \tfrac{1}{2}$	$\lambda_{Zn} = -\tfrac{1}{2}$	$\tfrac{1}{2}\ln[Zn^{2+}]$	$E_p^{\ominus} - \tfrac{1}{2}RTF^{-1}\ln[Zn]_{Hg}$
Metal/insoluble salt	$\boxed{AgCl} - \boxed{Ag} - Cl^- + \varepsilon = 0$	$\lambda_{Cl^-} = -1$	$\lambda_{AgCl} = 1$ $\lambda_{Ag} = -1$	$-\ln[Cl^-]$	$E_p^{\ominus}$
	$\tfrac{1}{2}\boxed{PbF_2} - \tfrac{1}{2}\boxed{Pb} - F^- + \varepsilon = 0$	$\lambda_{F^-} = -1$	$\lambda_{PbF_2} = \tfrac{1}{2}$ $\lambda_{Pb} = -\tfrac{1}{2}$	$-\ln[F^-]$	$E_p^{\ominus}$
'Redox' (bright platinum)	$Fe^{3+} - Fe^{2+} + \varepsilon = 0$	$\lambda_{Fe^{3+}} = 1$ $\lambda_{Fe^{2+}} = -1$		$\ln[Fe^{3+}]/[Fe^{2+}]$	$E_p^{\ominus}$
	$\tfrac{1}{2}Tl^{3+} - \tfrac{1}{2}Tl^+ + \varepsilon = 0$	$\lambda_{Tl^{3+}} = \tfrac{1}{2}$ $\lambda_{Tl^+} = -\tfrac{1}{2}$		$\tfrac{1}{2}\ln[Tl^{3+}]/[Tl^+]$	$E_p^{\ominus}$
	$\tfrac{1}{2}\boxed{Q} + H^+ - \tfrac{1}{2}\boxed{QH_2} + \varepsilon = 0$	$\lambda_{H^+} = 1$	$\lambda_Q = \tfrac{1}{2}$ $\lambda_{QH_2} = -\tfrac{1}{2}$	$\ln[H^+]$	$E_p^{\ominus}$

* boxes represent species in their standard state

reduced ones. The half-cell potential is given by the Nernst equation in the form

$$E_p = E_p^0 + RTF^{-1}\left\{\sum_I \lambda_I \ln [I] + \sum_U \lambda_U \ln [U]\right\} \qquad (8.9)$$

Pure solids and liquids are by convention in their standard state, and those uncharged species which are not in their standard state are usually at constant concentration during a particular experiment. Equation (8.9) can then be simplified to

$$E_p = E_p^{0'} + RTF^{-1}\sum_I \lambda_I \ln [I] \qquad (8.10)$$

where the term

$$E_p^{0'} = E_p^0 + RTF^{-1}\sum_U \lambda_U \ln [U] \qquad (8.11)$$

is a constant. Examples of half-reactions which form the bases of probe electrodes are shown in Table 8.1. The second column from the right shows the concentration-dependent part of the half-cell potential which, in all but two cases, involves only one species. If the value of E_p is indeed given by equation (8.9), the half-cells can therefore be used as probes for these species.

The Nernst equation (8.9) is, however, only applicable to reversible half-cells in which equilibrium at the electrode is established within the time of measurement and in which no reaction occurs to any appreciable extent. The half-cells shown in Table 8.1 are effectively reversible, although the hydrogen-gas electrode attains its reversible potential rather slowly, even at the catalytic surface of the platinum-black electrode. Substances such as lithium or fluorine, which react with water, cannot be incorporated into half-cells which contain aqueous solutions; nor can aluminium be used, as it acquires a coating of oxide which inhibits response to Al^{3+}.

'Redox' electrodes, such as Pt, Fe^{3+}, Fe^{2+} and Pt, Tl^{3+}, Tl^+, which each respond to the ratio of concentrations of the two participating ions, can of course only be used as a probe for one of them if the concentration of the other remains constant. But since the complexes formed by a given ligand with, for example, Fe^{3+} and Tl^{3+}, are often much more stable than the corresponding complexes of Fe^{2+} and Tl^+, it is often possible to use concentrations of ligand at which the ion of lower charge is effectively uncomplexed and is therefore of known concentration. A redox electrode may then be used as a probe for the ion of higher charge.

The symbols Q and QH_2 represent quinone(I) and hydroquinone(II), which are both sparingly soluble in water. A bright platinum electrode in contact with an aqueous suspension of the $1:1$ adduct $Q.QH_2$ ('quinhydrone') acts as an almost instantaneous probe for hydrogen ions.

O ----H——O

Q (I)
quinone

QH₂ (II)
hydroquinone

O ----H——O

quinhydrone

8.3   Probes based on membrane action

In this context, a 'membrane' is a phase which, when placed between two solutions, is permeable to some ions but not to others. For example, a crystal of silver chloride acts as a membrane if it is placed between two solutions either of silver nitrate or of sodium chloride. In the first case it is permeable only to silver ions, and in the second, only to chloride ions; and in either case a potential arises at the membrane. When we say that a membrane is permeable to a particular ionic species, we mean that ions of this type can both enter, and leave, the membrane phase. The same individual does not necessarily travel from one side of the membrane to another.

A membrane potential may be considered as a special case of diffusion potential (see sect. 8.4). At a liquid–liquid junction, all ions cross the boundary at rates which depend on their mobilities. However, when a membrane is placed at the boundary, the transport of some ions is prevented, i.e. these ions cross at zero rate.

We can try to quantify the potential E_B which arises at a boundary B between two solutions. Agar and Wagner have devised an elegant and versatile expression relating changes in the e.m.f. of a cell to gradients in the chemical potentials of its component species S. If activity coefficients are controlled, it can be written in the form

$$dE = RTF^{-1} \sum_S (\lambda_S - w_S) \, d \ln[S] \qquad (8.12)$$

The stoichiometric coefficients λ_S correspond to the terms λ_I and λ_U in equation (8.9) and the Washburn numbers

$$w_S = t_S/z_S \qquad (8.13)$$

represent the ratio of Hittorf transport number t_S to ionic charge z_S; they are negative for anions and zero for uncharged species. The e.m.f. of a cell may be obtained by integrating equation (8.12) over all regions of concentration gradient and summing the contributions of all species.

It is convenient, although thermodynamically improper, to use equation (8.12) to estimate the contribution of the boundary potential to the total e.m.f. The impropriety is due to the fact that boundary potentials, like single electrode potentials, can never be measured; experiment can only yield the potential difference between the terminals of the cell and can never allow us to break down this quantity into contributions from any particular part of the cell. With this caveat we shall attempt to calculate the potential E_B which arises at the boundary B in the arrangement:

$$C^{z+}A^{z-}(m_L) \underset{B}{\mid} C^{z+}A^{z-}(m_R) \tag{8.14}$$

As no electrode reaction occurs, $\lambda_C = \lambda_A = 0$ and equation (8.12) becomes

$$dE_B = -RTF^{-1}(w_C\, d\ln[C] + w_A\, d\ln[A]) \tag{8.15}$$

and since $[C] = [A]$, we have

$$dE_B = -RTF^{-1}(w_C + w_A)\, d\ln[C] \tag{8.16}$$

Integration across the boundary gives

$$E_B = -RTF^{-1}(w_C + w_A)\ln\frac{m_R}{m_L} \tag{8.17}$$

Now

$$w_C = t_C/z \tag{8.18}$$

and

$$w_A = -t_A/z \tag{8.19}$$

whence

$$E_B = RTF^{-1}\left(\frac{t_A - t_C}{z}\right)\ln\frac{m_R}{m_L} \tag{8.20}$$

At a liquid–liquid junction, the values of w_C and w_A (and hence of t_C and t_A) depend on the mobilities in aqueous solution. However, at a membrane, the values of w and t for one of the ions will be zero, provided that the membrane behaves ideally. Thus, if a membrane is impermeable to anions, $t_A = 0$, $t_C = 1$ and the potential E_{MC} at the membrane is given by equation (8.20) in the form

$$E_{MC} = -\frac{1}{z}RTF^{-1}\ln\frac{[C]_R}{[C]_L} \tag{8.21}$$

Similarly, the potential E_{MA} of a membrane which is permeable only to A^{z-} is given by

$$E_{MA} = \frac{1}{z}RTF^{-1}\ln\frac{[A]_R}{[A]_L} \tag{8.22}$$

Continuing along our non-rigorous path, we may calculate the potential across the set-up

---

$$-C^{z+} A^{z-} \quad\bigg|\quad -C^{z+} A^{z-} \quad\bigg|\quad \text{probe for } A^{z-} + \qquad (8.23)$$

$$m_L \qquad\qquad\qquad m_R$$

$$\qquad (2) \qquad\qquad\qquad (1)$$

$$\text{membrane}$$
$$\text{permeable}$$
$$\text{to } C^{z+}$$

---

where the electrode half-reaction at the probe for A^{z-} may be written

$$\frac{1}{z} A^{z-} + \sum_U \lambda_U [U] + \varepsilon \rightarrow 0 \qquad (8.24)$$

(cf. equation (8.8)). The e.m.f. across device (8.23) is the sum of the potentials $E_{(1)}$ and $E_{(2)}$ across the interfaces (1) and (2). Now from equation (8.10)

$$E_{(1)} = \text{constant} + RTF^{-1} \ln [A]_R^{1/z} \qquad (8.25)$$

and, from equation (8.21), the membrane potential $E_{(2)}$ is given by

$$E_{(2)} = -\frac{1}{z} RTF^{-1} \ln \frac{[C]_R}{[C]_L} \qquad (8.26)$$

If the composition of the right-hand solution is kept constant, we may write

$$E_{(8.23)} = E_{(1)} + E_{(2)} = \text{constant} + \frac{1}{z} RTF^{-1} \ln [C]_L \qquad (8.27)$$

Equation (8.27) is formally similar to the expression (8.11) for the potential of a probe electrode: and a device consisting of a membrane permeable to one ion, I, a solution of constant composition containing that ion, and a probe which is reversible to the counter-ion, can indeed be used as a probe for the ion I.

Equations (8.21) to (8.27) are based on the assumption that membranes behave ideally; but in practice, few, if any, membranes are permeable to only one type of ion. The oldest and most familiar membrane electrode, the glass electrode (see Fig. 8.1(*b*)), is not strictly permeable to any ion although it responds in some way to almost any positively charged one. A thin layer of glass between two aqueous solutions behaves as if it were three distinct regions: two wet phases W_R and W_L in contact with the solutions to the right and left of it, and a central dry region D. Cation exchange occurs between each wet region and the adjoining solution; and solid–solid junction potentials are set up, within the glass, each side of the dry region. The potential across the glass is the sum of the potentials at the boundaries W_R–D and

D–W_L for all ions. For any one ion (e.g. H^+) at any one junction (e.g. W_R–D) we may write

$$E_{j(W_R-D)} = -RTF^{-1} \ln \frac{[H]_{W_R}}{[H]_D} + d_R \tag{8.28}$$

$$= -RTF^{-1} \ln \frac{[H]_R}{[H]_D} + d_R' \tag{8.29}$$

where d_R and d_R' are constants. The contribution of the hydrogen ion to the total potential across the glass is then

$$E_{MH} = -RTF^{-1} \ln \frac{[H]_R}{[H]_L} + d \tag{8.30}$$

indicating that, if there were no contribution from other cations, a glass membrane could be incorporated into a device such as (8.23) to act as a probe for hydrogen ions. The 'asymmetry potential' d depends on the detailed structure of the wet glass surfaces. Although it varies somewhat with time, it can usually be trusted to remain effectively constant within the periods needed to make measurements.

If a membrane responds to more than one type of cation, each species will contribute to the potential by a term which depends on its concentration, its partition coefficient P between wet glass and aqueous solution, and its mobility u within the wet glass. The values of hydrogen ion concentration in equations such as (8.28) and (8.29) must therefore be replaced by a term $\sum [C]'$ which contains the weighted concentrations of all cations present. If the glass responds to any monovalent cation this term would be given by

$$\sum [C]' = [H^+] + \sum K_M[M^+] \tag{8.31}$$

where the selectivity coefficient

$$K_M = P_M u_M / P_H u_H \tag{8.32}$$

is very sensitive to composition and for some glasses is as low as 10^{-15} for any cation other than H^+. Similar selectivity coefficients may be used to describe the response of membranes to anions.

The glass membrane is the most frequently used membrane electrode; indeed, it is the most frequently used probe of any sort. But other membrane electrodes are being developed (see Fig. 8.2). They include the single crystal LaF_3 membrane, used as a probe for fluoride ions (p. 96). Membranes may also be made by setting the active substance in a matrix, e.g. silver halides and some transition metal sulphides are prepared in a matrix of silver sulphide or of silicone rubber. The components written to the right of the membrane in equation (8.23) are often contained inside a vessel of which the membrane forms the base. Sometimes, as with glass electrodes, they are sealed into the container. 'Membrane' electrodes may, however, also be prepared by depositing the active material on to a base of graphite; despite the absence of the right-hand part of the cell, such electrodes show a Nernst

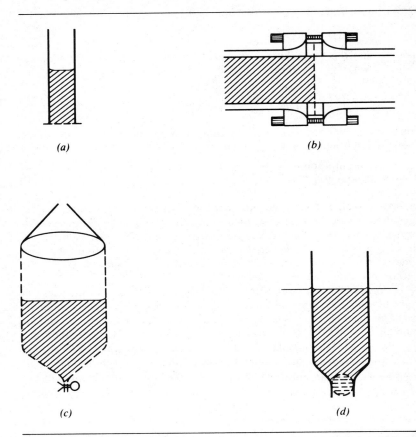

(a)   *(b)*

(c)   *(d)*

Fig. 8.2 Some membrane electrodes; the permselective material is represented by a dotted line. (*a*) membrane or crystal cemented to end of tube; (*b*) sheet membrane clamped between flanged glass tubes; (*c*) cylindrical membrane cast over a test-tube and sealed at one end; (*d*) plug of ion-exchange resin. (After G. V. Hills, in D. J. G. Ives and G. J. Janz, eds., *Reference Electrodes*, Academic Press, New York (1961).)

response similar to that of conventional membrane electrodes. Attempts have also been made to develop liquid membranes, which consist of a solution of a reagent for metal ions in a solvent which is immiscible with water; but they are less selective than solid membranes and so are less suitable for use as probes for studies of equilibria.

8.4   The liquid junction

The value of the potential E_J which arises from free diffusion of ions can also be obtained from the Agar–Wagner equation (8.12), with all values of λ_S

equal to zero. Thus

$$dE_J = -RTF^{-1} \sum w_I \, d \ln [I] \tag{8.33}$$

Suppose that the two solutions both contain two symmetrical electrolytes $B^{z+}A^{z-}$ and $D^{y+}C^{y-}$ of concentrations m and m' respectively. Then equation (8.12) becomes

$$E_J = RTF^{-1}\left[\left(\frac{t_A - t_B}{z}\right)\ln\frac{m_R}{m_L} + \left(\frac{t_C - t_D}{y}\right)\ln\frac{m'_R}{m'_L}\right] \tag{8.34}$$

from which it follows that the values of E_J can be kept near to zero if:

(i) The concentrations of the two solutions are as similar as possible so that the ratios m_R/m_L and m'_R/m'_L approach unity.
(ii) The component ions of the two electrolytes have similar mobilities. Then the differences $(t_A - t_B)$ and $(t_C - t_D)$ approach zero.
(iii) The concentration of one electrolyte is much larger than the other and is the same in both solutions.

If the concentration of a bulk electrolyte B^+A^- is m and the mean concentration of the variable component D^+C^- in the two solutions is $\bar{m}'$, the approximate value of the junction potential is given by

$$E_J = RTF^{-1}\frac{\bar{m}'(u_C - u_D)}{\delta m(u_B + u_A) + \bar{m}'(u_D + u_C)}\ln\frac{m'_R}{m'_L} \tag{8.35}$$

where u_S is the ionic mobility of the species S. So the junction potential between two ionic solutions containing the same ionic medium but different, much lower, concentrations of a second ionic solute can be decreased by increasing concentration of bulk electrolyte. In practice, the only solutes which contribute markedly to E_J at concentrations less than ~ 0.1M are those in which the mobility of one ion is much higher than the other, viz. strong acids and to a lesser extent, strong alkalis. It is found that E_J is proportional to $[H^+]$ or $[OH^-]$, the proportionality constant for each ion decreasing with increasing concentration of bulk electrolyte and, for a given medium, being greater for H^+ than for OH^-. Studies of weak complexes, such as acetates, may be complicated by the need to use such high concentrations of ligand that its contribution to E_J cannot be ignored.

8.5 The choice of probe

With so wide a variety of probe available which one should we choose? Obviously, the first decision is which ion(s) to monitor. The vast majority of studies of equilibria have involved the measurement only of pH and related quantities, usually by means of a glass-membrane electrode; and since most ligands are basic, it seems likely that this method will continue to be popular.

However, there are other probes for H^+; and for precise work it may be desirable to use a probe for the free metal ion. The concentrations of some simple ligands, both basic (e.g. F^-) and non-basic (e.g. Cl^-), may also be measured directly. The choice of probe hinges on such diverse criteria as the precision which can be obtained, the time and manual skill needed, and the type of equipment available.

Hydrogen ion probes

There are three main types:

1. The hydrogen-gas electrode has the appeal of intellectual simplicity and low electrical resistance (see below), but it needs a cylinder of hydrogen and a lot of time. The electrode takes at least one hour to reach equilibrium; and this makes a titration protracted. Substances which are reduced by gaseous hydrogen or which poison the catalytic surface of the platinum-black electrode interfere.

2. The quinhydrone electrode (see sect. 8.2) is much more convenient and comes to equilibrium almost instantaneously, but its use is restricted to solutions of pH between 1 and 7. At higher acidities, Q becomes protonated; and in alkaline solutions, QH_2 becomes deprotonated. Some metal ions, such as Cu^{2+}, interfere by displacing a proton from QH_2 to form a complex $CuQH^+$. The quinhydrone electrode, like the hydrogen-gas electrode, has the advantage of a very low electrical resistance. The e.m.f. of cells containing either electrode can therefore be measured very precisely (to $\pm 0 \cdot 01$ mV) with a null-point potentiometer under conditions of strict reversibility where no current is being drawn from the cell (see sect. 8.8).

3. The glass electrode. A typical glass membrane electrode, including the sealed-in right-hand half-cell of the device (8.23) is shown in Fig. 8.1(*b*). The right-hand electrode (or 'internal reference electrode') is usually AgCl, Ag and the internal electrolyte is aqueous HCl, which provides H^+ and Cl^- ions at constant concentration in contact with the inner surface of the glass.

Glass electrodes are of much higher electrical resistance than other probes for H^+, and so can only be used in conjunction with a 'pH-meter' or other potentiometer which incorporates an amplifying device. The usual precision is about $\pm 0 \cdot 2$ mV. With proper care, however, measurements can be made to $\pm 0 \cdot 01$ mV, a precision sufficient to unscramble quite complicated multiple equilibria and to produce work of very high quality. Some glass electrodes give a Nernst response to H^+ over a very wide range of pH, e.g. $1 \leqslant pH \leqslant 13$. They come to equilibrium rapidly and are unaffected by those substances which interfere with hydrogen and quinhydrone electrodes. Their main disadvantage, paradoxically, is the great sensitivity of the glass membrane. This must, of course, be protected from even slight mechanical damage; but the structure is also influenced by changes of temperature and solvent and after any such disturbance the 'constant' value of the asymmetry potential d in equation (8.30) may drift much more markedly than usual, sometimes for several days. The glass surface may be incapacitated by the absorption of polyelectrolytes or of small, highly charged cations. However, with thoughtful handling and strict observance of the maker's instructions, a glass electrode often proves to be an invaluable probe which may survive many months, or even years, of constant use.

Metal ion probes

Metal ions, like hydrogen ions, can be monitored by means of several types of probe:

(i) *Metal or amalgam electrodes* The majority of metals which undergo reversible electron transfer with their aqueous cations are in the B subgroups of the periodic table. Most other metals cannot be used in this way because they either react rapidly with water (e.g. those in groups IA and IIA) or (like aluminium and iron) become passive by acquiring a surface film of oxide. Those which do form reversible half-cells are often used as the amalgam rather than the pure metal. Amalgams come to equilibrium more quickly; but they need skilled handling because rapid oxidation may occur unless air is excluded. Being electronic conductors, both metals and amalgams have the advantage of very low electrical resistance.

(ii) *Ion-selective membrane electrodes* Reactive metals, such as those in groups IA and IIA, may often be monitored using high-resistance electrodes of glass and other membranes (see sect. 8.3). Although the selectivity of these electrodes is still rather low, the field is developing fast, and the manufacturers' latest specifications should be consulted.

(iii) *Redox electrodes* A platinum flag in contact with the ions M^{z+} and M^{y+} responds rapidly to the ratio of the concentrations of the two oxidation states; and if one of the states is unaffected by the presence of a complexing agent, a redox electrode may be used as a probe for the other state. The hydrolysis of Fe(III) has, for example, been studied using the redox electrode Fe(III)/(II) in solutions of pH < 3 where the iron(II) is almost completely unhydrolysed ($\bar{n} < 10^{-6}$). Although redox electrodes are of low resistance and very convenient, the presence of any 'indicator species' (in this case iron(II)) introduces an extra hazard in interpretation. They should therefore be used with discretion.

Probes for anions

Studies of equilibria seldom involve direct monitoring of anions, since most are basic and are traditionally observed by the pH method. The concentrations of halide and sulphate ions are, however, often measured using a metal, insoluble salt electrode, such as AgCl (Ag + Cl$^-$). These electrodes are of low resistance, but are not always free from interference troubles. The response of the silver chloride electrode is, for example, affected by bromide ions. Since $[OH] = K_w[H]^{-1}$, any probe for hydrogen ions may also be calibrated for use as a probe for hydroxyl ions.

The range of anion probes has been much widened by the development of ion-selective membrane electrodes, e.g. for halide, phosphate, sulphide and sulphite ions. But, as with membrane electrodes for metal ions, many are as yet insufficiently selective for precise studies of equilibria. The best one so far is a single crystal of LaF_3, doped with europium to increase the mobility of the fluoride ion (see Fig. 8.2). The membrane gives a Nernstian response to fluoride ion activities between 1M and 10^{-5}M provided that $[OH^-] < 10^{-5}$M. There is even less interference from other anions. The LaF_3 electrode has been widely used to study fluoride complexes and is a great advance on the previous potentiometric (redox) technique, in which two indicator species were added: iron(II) and iron(III). The potential of a platinum flag gave the ratio $[Fe^{3+}]/[Fe^{2+}]$ from which the value of $[F^-]$ could be calculated, provided that the stability constants of iron(III) and iron(II) fluoride complexes had been measured under identical conditions.

8.6   The reference half-cell and junction

Almost all potentiometric studies of stability constants have been carried out using cells which contain liquid junctions. The great majority of these cells incorporate the glass and calomel electrodes supplied with a commercial pH-meter. The e.m.f. of the calomel reference half-cell

$$Hg(1) | Hg_2Cl_2(s) | Cl^- \text{ (sat aq KCl)} \tag{8.36}$$

is constant at a given temperature. The saturated aqueous solution of KCl also acts as a salt bridge between the reference half-cell and the test solution (see Fig. 8.1(a)) and, since $t_K \simeq t_{Cl}$, the junction potential is minimised (and probably negligible unless the test solution is of high acidity or is only partly aqueous). A saturated KCl bridge may, however, cause trouble in studies of metal ions which combine appreciably with chloride ions because there will be some unavoidable leakage of chloride ions from the bridge into the test solution; and it is certainly not to be recommended for studies of metal ions such as Ag^+ which form insoluble chlorides.

It is better to use the same ionic medium as background electrolyte for both test solution and reference half-cell and to use the same medium, without other solutes, as bridge electrolyte. A convenient arrangement for normal use is shown in Fig. 8.3. Any reliable half-cell may be used as reference electrode. Two popular ones, suitable for studying equilibria in 3M sodium perchlorate, are

$$\begin{array}{c|c|c|c|c} Ag(s) & AgCl(s) & NaCl & NaClO_4 & NaClO_4 \\ & & (aq\ 0{\cdot}010M) & (aq\ 2{\cdot}990M) & (aq\ 3{\cdot}000M)\ bridge \end{array} \tag{8.37}$$

and

$$\begin{array}{c|c|c|c} Ag(s) & AgClO_4 & NaClO_4 & NaClO_4 \\ & (aq\ 0{\cdot}010M) & (aq\ 2.990M) & (aq\ 3{\cdot}000M)\ bridge \end{array} \tag{8.38}$$

Titration assemblies have been designed so that cells such as (8.37) and (8.38) may be used with small volumes of solutions (2 to 10 ml) inside a glove-box.

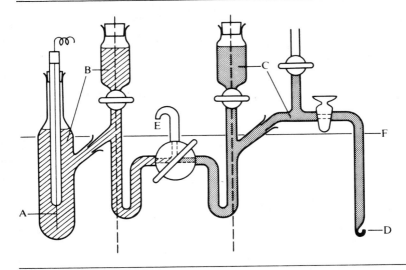

Fig. 8.3 Reference half-cell and liquid junction (e.g. for (8.37) and (8.38)). The assembly may conveniently be bent, perpendicularly to the paper, along the dotted lines. A. Electrode of reference half-cell. B. Solution of reference half-cell. C. Bridge solution. D. Junction between test- and bridge-solutions. E. To waste. F. Level of thermostat bath. (After F. J. C. Rossotti and H. Rossotti, *The Determination of Stability Constants*, McGraw-Hill, New York (1961).)

8.7   Cells without liquid junction

In order to increase the rigour of the work by eliminating the liquid junction, two types of cell have occasionally been used:

1. The Harned cell:

$$-\,\text{Ag(s)}\,\Big|\,\text{AgCl(s)}\,\Big|\,{\text{test solution}\atop \text{H}^+,\,\text{Cl}^-,\,\text{M}^{z+},\,\text{L}}\,\Big|\,{\text{H}_2(\text{g})}\,\Big|\,{\text{Pt}+\atop(\text{black})} \qquad (8.39)$$

for which

$$E = E^0 + RTF^{-1}(\ln\{\text{H}^+\}\{\text{Cl}^-\} - \tfrac{1}{2}\ln p_{\text{H}_2}) \qquad (8.40)$$

has been used to study complex formation between a basic ligand L and metal ions M^{z+} which do not complex appreciably with chloride ions. If multiple equilibria occur, interpretation is very difficult unless a bulk electrolyte is used. Then, if $[\text{Cl}^-]$ is known, the e.m.f. of the Harned cell gives $[\text{H}^+]$, from which the free ligand concentration can be obtained provided that the protonation constants of L have been measured under identical conditions. The procedure is exactly analogous to that described in sect. 8.1 except that uncertainties about the junction potential are now replaced by uncertainties about interaction between M^{z+} and Cl^-. The latter seems worse.

   2. A cell such as

$$-Ag(s) \mid AgCl(s) \mid \begin{array}{l} \text{test solution} \\ M^{z+}, Cl^- \end{array} \mid M(s) + \qquad (8.41)$$

may contain two probe electrodes, one for metal ion and one for ligand (in this case, Cl^-). Since the e.m.f. of cell (8.41) is given by

$$E = E^0 + RTF^{-1} \left\{ \frac{1}{z} \ln [M] + \ln [L] \right\} \qquad (8.42)$$

the value of

$$\ln \alpha_0 [L]^z = \frac{zF(E - E^0)}{RT} - \ln M_t \qquad (8.43)$$

may be calculated. The value of $\alpha_0[L]^z$ is, of course, a polynomial in [L] only, and so the stability constants may be obtained from the variation of E with M_t and L_t by successive approximations. But this inconvenient method is of very limited use.

8.8   Measurements

We shall discuss the potentiometric measurement of concentration, using cells with liquid junction which respond to the concentration of a single species S. If the cell behaves reversibly, the e.m.f. is given by equation (8.4) which predicts a linear relationship between E and $\ln [S]$ provided that E_J is negligible. The intercept E^0 of the function $E(\ln [S])$ may then be determined from solution in which [S] is known, and serves to calibrate the cell as a probe for S.

   Reversible measurements can be made only if a negligible current is drawn from the cell. This ideally implies using a null-point potentiometer. Some cells, however (e.g. those containing glass electrodes), have too high a resistance to be studied in this way and can only be used in conjunction with a 'pH-meter', which acts by amplifying the off-balance current; but the amount of current actually drawn from the cell is insufficient to produce any observable irreversibility.

   Reversibility may be checked in two ways. Micropolarisation tests involve drawing a small current I from the cell and impressing a small voltage V across it. A linear plot of I against V (rather than a hysteresis loop) implies reversibility. The validity of the Nernst equation should always be tested directly by checking that E is a linear function of the natural logarithm of the appropriate concentration term (see equation (8.4)) and that the proper gradient of $\lambda_S RTF^{-1}$ is obtained. The temperature must, of course, be carefully controlled.

The way in which the current flowing through a polarised cell varies with the applied voltage can also give information about complex formation. But the stability constants obtained from current–voltage plots (or polarograms) are usually less precise than those obtained from e.m.f. measurements. Since polarography has few, if any, compensating advantages over potentiometry, it is a much less useful technique for the equilibrium chemist and will not be discussed further.

As an illustration of the potentiometric method, we shall outline a convenient procedure for measuring the stability constants of complexes ML_n by means of a cell such as (8.1) containing a glass electrode and a reference half-cell and junction. The titrations are planned so that the glass electrode remains immersed in test solution throughout each set of measurements. We assume that the stock solutions are made up with the appropriate salt background and that values of H_t, M_t and L_t are known. We shall follow the most popular procedure of starting with a test solution of low pH and titrating with base to reduce the value of H_t. If measurements are required only in the acidic range (pH < 6) it is often better to use $NaHCO_3$ rather than strong base as titrant, to avoid local precipitation of metal hydroxide. Table 8.2 shows stock solutions which would be suitable for measurements in a 3M $(Na)ClO_4$ medium.

Table 8.2 Convenient stock solutions for potentiometric titrations.

		$[Na^+]$	$[ClO_4^-]$	
Acid	$[H^+] = H_i$	$3 - H_i$	3	M
Base	$[HCO_3^-] = B_i$	$3 + B_i$	3	M
Metal ion	$[M^{z+}] = M_i [H^+] = H_t'$	$3 - zM_i - H_i'$	3	M
Ligand (as H_jL)	$[H_jL] = L_t$	3	3	M

The titration is best carried out in several stages:

1. An aliquot of acid solution is titrated with base, and the results plotted in the form of a Gran function (see sect. 7.6) in order to check that the equivalence point obtained from the plot is the same as that predicted from the values of H_i and B_i. It is not necessary that the titrations should actually reach the equivalence point, nor that the value of E^0 be known at this stage. The value of $(E - RTF^{-1} \ln [H])$ is calculated and plotted against [H]. A constant value $(= E^0)$ implies both that the cell is behaving reversibly and that activity coefficients are constant. If the value of $(E - RTF^{-1} \ln [H]) = E^{0'}$ departs from E^0 at high acidities, the difference between the two quantities may be ascribed to an appreciable junction potential E_J and may well be proportional to [H]. Once the proportionality constant $E_J/[H]$ has been determined, the value of [H] in a solution of unknown acidity can always be obtained from values of E and $E^{0'}$ by successive approximations.

2. The protonation constants of the ligand may be determined by adding an aliquot of ligand solution, together with a further aliquot of acid to ensure that $\bar{j}$ is as high as possible, and then titrating with base. Alternatively, the ligand solution alone may be added to the nearly neutral solution obtained by the end of stage 1 and titrating this with acid. The value of [H] is calculated from the values of E and $E^{0\prime}$ and combined with values of H_t and L_t (calculated from H_i, B_i, L_i and the volumes of stock solutions used) to give the function $\bar{j}([H])$, and hence the protonation constants.

3. An aliquot of metal ion solution is then added, together with yet more of the stock solution of acid in order to ensure that all, or most, of the ligand is protonated, rather than complexed with metal ion. The solution is again titrated with base and values of [H] as a function of H_t, L_t and M_t are obtained.

Although it must obviously be checked that the measurements are reproducible, exact repetition of any particular titration is probably a waste of time. It is better to vary the initial concentrations of ligand and metal ion. If identical functions $\bar{j}([H])$ and $\bar{n}([L])$ are obtained using different concentrations of reactants, the measurements are clearly reproducible and moreover, no appreciable concentrations of mixed or polynuclear complexes are formed (see Ch. 11). Repetition of titrations is probably only necessary as one step in trouble-shooting if the formation functions obtained from different measurements on the same system do not coincide.

The procedure outlined above can, of course, be varied greatly. When satisfactory values of the protonation constants have been obtained for two or three different concentrations of ligand, stage 2 may be omitted, or replaced by the titration with base of a mixture of solutions of metal ion and acid, in order to check, by Gran plot, the total concentration of acid in the metal ion solution. The ligand solution can be added at the end of this titration. Some workers like to keep the total concentration of metal ion constant throughout a titration and this can readily be done by adding aliquots of metal ion solution at the same time as titrating with base. This procedure is almost essential if polynuclear complexes are formed (see sect. 11.1), but seems unnecessary for a purely mononuclear system. The detailed tactics must, of course, be varied to suit the basicity of the ligand, the stability of the complexes and the solubilities of both. But it is clear that for any multistage titration of the sort outlined above, the titration vessel must be designed so that the electrodes can be covered by a very small proportion of the total volume of liquid which can be contained in the cell.

Potentiometry is not only the most precise method for determining stability constants; it is also the most widely applicable. The function $\bar{n}([L])$ can be obtained for complexes of any basic ligand provided that the values of L_t can be high enough to provide a sufficient free ligand concentration to form complexes and that M_t can be high enough to ensure that complex formation results in a measurable charge in [L]. Complexes of a few non-basic ligands may be similarly studied. Probes are also available for a number of metal ions, and lead to values of α_0. The value of [L] is often determined in the same titration using a glass electrode; but, if necessary, [L] may be obtained from α_0 and L_t by successive approximations.

Two types of labile system cannot normally be studied potentiometrically: those in which M (e.g. an actinide ion) is available only at tracer concentrations, and those in which one or more species are insufficiently soluble in the medium used. Methods for dealing with complexes of this type are discussed in Chapter 9. However, with its high precision and wide range, potentiometry should surely be the preferred technique for determining stability constants in any system to which it can be applied.

Experimental methods: II. Two-phase systems

Measurement of the distribution of either metal ion or ligand between two phases can give information about both the partition equilibria between the phases and the homogeneous equilibria which are set up within one or both of the phases. The distribution of ligand between aqueous solution and vapour was used in Jannik Bjerrum's classic study of ammonia complexes of metal ions. Measurement of the vapour pressure of ammonia over the aqueous solutions gave the concentration of free ligand, and hence $\bar{n}$. But two-phase studies of metal complex formation in aqueous solution more often involve measurement of the distribution of metal ion between the solution and a second phase, which may be a sparingly soluble salt of the metal, an immiscible organic solvent which extracts the uncharged metal complex or, less frequently, a cation exchange resin.

These three methods will be briefly discussed in turn.

9.1 Solubility

The use of solubility measurements in studies of metal complex formation have been reviewed by Johannson. The simplest application of the method involves equilibrating a sparingly soluble salt ML_c with a solution containing the ligand L and sufficient bulk electrolyte to control the activity coefficients in solution. If the composition and structure of the solid remain the same, so too do the activities of the component ions. The stoichiometric solubility product $K_s = [M][L]^c$ of ML_c is also a constant which is, of course, the equilibrium constant for the partition of the ionic compound $(M^{z+} + cL^{(z/c)-})$ between the pure solid and the ionic medium. Since the solubility, S, of ML_c gives the total concentration of metal ion, we have

$$S = M_t = \sum_0^N [ML_n] \tag{9.1}$$

$$= [M] \sum_0^N \beta_n [L]^n \tag{9.2}$$

$$= K_s \sum_0^N \beta_n [L]^{n-c} \tag{9.3}$$

$$= K_s [L]^{-c} \alpha_0^{-1} \tag{9.4}$$

The solubility S is therefore a function of [L] only, and the parameters K_s and β_n may be obtained from the measurements S, [L]. There are two main disadvantages in this method: (1) we need to introduce an extra parameter, K_s, in addition to the stability constants; (2) as the solutions are in equilibrium with solid ML_c, we cannot vary [M] and [L] independently.

Equation (9.4) accounts for the fact that the solubility passes through a minimum as [L] increases. Figure 9.1 shows the variation of solubility of

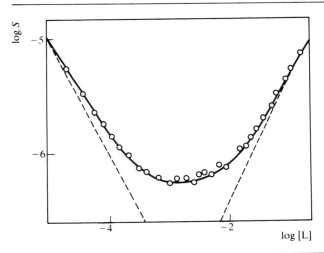

Fig. 9.1 Log S as a function of log [L] for the solubility of AgCl in solutions containing Cl^- in 9·88 per cent aqueous ethanol. The dotted lines are of slope −1 and +1. (After K. P. Anderson, E. A. Butler, D. R. Anderson and E. M. Woolley, *J. Phys. Chem.*, **71**, 3566 (1967).)

AgCl with chloride ion concentration in aqueous ethanol. Here

$$S = [Ag^+] + [AgCl] + [AgCl_2^-]$$
$$= K_s\{[Cl^-]^{-1} + \beta_1 + \beta_2[Cl^-]\} \qquad (9.5)$$

so that, as $[Cl^-]$ is raised, S first decreases (the so-called 'common ion' effect) and then increases on account of the formation of $AgCl_2^-$ and any higher complexes. The free ligand concentration is, strictly,

$$[Cl^-] = [Cl^-]_i + (1 - \bar{n})S \qquad (9.6)$$

but when the value of S is very low, the value of $[Cl^-]$ is effectively equal to the chloride ion concentration $[Cl^-]_i$ in the initial solution.

At low free ligand concentrations, where the plot of log S against log $[Cl^-]$ has a slope of −1, the silver is present in solution largely as Ag^+. At high concentrations of chloride the slope, d log S/d log $[Cl^-]$, approximates to +1, indicating that $AgCl_2^-$ is predominant form of silver in solution. Some less simple solubility curves are shown in Fig. 9.2. In the lutetium oxalate system, (*a*), the insoluble salt is $M_2L_3.7H_2O$ and the highest complex formed is ML_4. The europium oxalate system, (*b*), is further complicated by the (slow) transition between the two solid phases $2NaML_2.7H_2O$ and $M_2L_3.10H_2O$, the latter being the more stable at lower concentrations of free ligand.

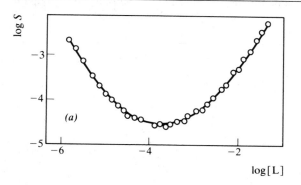

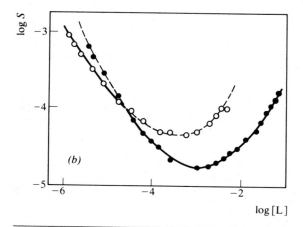

Fig. 9.2   Log S as a function of log [L] for the solubility of some lanthanide oxalates in aqueous oxalate solutions. The solid phases are: (*a*) $Lu_2ox_2.7H_2O$; (*b*) $NaEuox_2.7H_2O$ (open circles) and $Eu_2ox_3.10H_2O$ (full circles). The dotted curves denote solubility of a metastable phase. (From I. Grenthe and G. Gardhammar, *Acta Chem. Scand.*, **23**, 93 (1969).)

The sparingly soluble salt used for determining stability constants need not have both the metal ion and ligand in common with the complexes being studied. A salt of the same metal with an auxiliary ligand (often iodate) may be preferable. The solubility of MX_x in solutions containing L is given by

$$S = M_t = \sum_0^N [ML_n] = [M] \sum \beta_n [L]^n \tag{9.7}$$

provided that no complexes of M and X are present in solution. Then, so long as MX_x is the only solid present we have

$$[M] = K_s[X]^{-x} \tag{9.8}$$

and

$$[X] = xM_t = xS \tag{9.9}$$

Equation (9.3) may then be written as

$$S = K_s(xS)^{-x} \sum_0^N \beta_n[L]^n \tag{9.10}$$

or

$$x^x S^{x+1} = K_s \sum_0^N \beta_n[L]^n = K_s \alpha_0^{-1} \tag{9.11}$$

to give a familiar polynomial in [L], from which we can obtain values of K_s and β_n.

Obviously, if either the ligand L or the anion X is basic, the pH of the saturated solution must be measured so that allowance may be made for the concentrations of any protonated species calculated from acidity constants obtained under identical conditions.

The use of an auxiliary ligand extends the range considerably, since any salt MX_x with a convenient solubility may be used. The increase in the solubility of several metal iodates with increasing concentration of, for example, halide ions, has been used to calculate stability constants for halide complexes.

Solubility is, however, very tedious to measure, and even the most careful work lacks the precision of good potentiometry. Equilibrium is attained only slowly and the function $S([L])$ has to be built up by amassing 'single-point' measurements, rather than by titration. Each point is determined at constant temperature either by shaking the solution with the solid until it becomes saturated, or, better, by letting the solution seep through a Brönsted–Davies saturating column. The saturated solution is carefully freed from solid by filtration or centrifugation (preferably at the equilibration temperature) and then analysed for the total concentration of metal ion or auxiliary ligand. Solubilities which are not too low may be measured by traditional gravimetric or volumetric methods. Iodate, for example, may be measured iodimetrically. Lower solubilities may be determined by trace analysis, e.g. using polarography, colorimetry or radioactive tracers.

Although the solubility method is time-consuming and inconvenient, it does, with care, give reliable results and is useful for investigating systems in which neither $\bar{n}$ nor α_0 can be measured potentiometrically. For example, the difference in solubility of $Ca(IO_3)_2$ in aqueous 3M $NaClO_4$ and in a 3M $(Na^+)ClO_4^- - Cl^-$ mixture can be attributed to the formation of chloro complexes of Ca^{2+}. Since these are weak, the value of $\bar{n}M_t$ is low and $[Cl^-]$ differs so little from Cl_t that it is difficult to obtain a value of $'\bar{n}$.

Potentiometric determination of α_0 is also difficult in the absence of a reliable ion-selective electrode for calcium and for such systems solubility can provide information which cannot readily be óbtained in other ways.

The solubility method is being used increasingly frequently to study equilibria in solvents other than water including: mixed aqueous–organic solvents. (cf. Fig. 9.1); non-aqueous ionising solvents such as dimethyl-sulphoxide; and fused salts.

9.2   Solvent extraction

This technique also involves the distribution of the metal ion between two phases: an aqueous phase and a liquid organic phase which is immiscible with it. Activity coefficients in the aqueous phase are again controlled by the presence of a bulk electrolyte. Although activity coefficients cannot be similarly controlled in the organic phase, they do not often vary much with the concentration of metal ion extracted, because the species which partitions is normally uncharged. Activity coefficients in the non-aqueous phase are more sensitive to metal ion concentration when ion-pairs are extracted, e.g. into ethers. More drastic changes in activity coefficients occur, in both phases, when the composition of the two solvents varies with the concentration of ligand, as happens when, e.g., iron(III) is extracted into ether from concentrated hydrochloric acid. Such gross changes in the composition of the phases are often accompanied by changes in volume.

We shall optimistically assume that the activity coefficients in each phase are independent of composition. The stoichiometric stability constants are then constant, as is the partition coefficient $P_c = [ML_c]_0/[ML_c]$ where $[ML_c]_0$ and $[ML_c]$ represent the concentrations of the uncharged complex ML_c in the organic and aqueous phases.

The simplest application of solvent extraction to the determination of stability constants involves measuring the total concentration of metal ion in each phase in the presence of only one type of ligand (see Fig. 9.3). The ratio of these total concentrations is the distribution ratio, and is given by

$$q = \frac{[M_t]_0}{M_t} = \frac{[ML_c]_0}{\sum\limits_0^N [ML_n]} \tag{9.12}$$

$$= \frac{P_c[ML_c]}{\sum\limits_0^N [ML_n]} \tag{9.13}$$

$$= P_c\alpha_c \tag{9.14}$$

The distribution ratio is therefore a function of [L] only, and so measurements of q, [L] can yield values of the stability constants. As with the solubility method, the use of a second phase makes it necessary to introduce an extra parameter to describe the results.

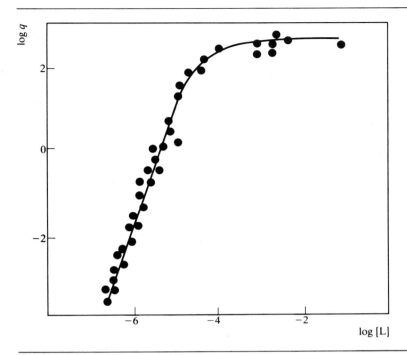

Fig. 9.3 Log q as a function of log [L] for the distribution of ^{233}Pa(IV) between benzene and aqueous acetonylacetonate solutions. (After R. Lundquist and J. Rydberg, *Acta Chem. Scand.*, A**28**, 243 (1974).)

When, as is often the case, the ligand is an organic anion, the uncharged acid H_cL may itself be extracted into the organic phase. In order to calculate the free ligand concentration in the aqueous phase, we always need to know its protonation constants and the equilibrium value of [H] in the aqueous phase; and if H_cL partitions we also need to know its partition coefficients (and, of course, the volumes of the two phases).

Equilibration and phase separation are quicker and easier for solvent extraction than for solubility. The distribution ratio of metal is usually determined using radioactive tracers, although spectrophotometric analysis for the metal may also be used. Values of q between 10^3 and 10^{-3} may be obtained with acceptable precision, and only very low concentrations of metal ion need be used. The technique is therefore well suited to the study of actinide complexes.

Radiochemical assay is best carried out by liquid-counting of the initial solution, and of both equilibrated phases. For a given counting assembly, the number of counts per minute is given by

$$N = k_s M_t \tag{9.15}$$

Since the constant k_s depends on the solvent as well as on the apparatus, the value of q cannot be obtained directly as the ratio of counting rates in the two phases. But the ratio k_{s1}/k_{s2} for two solvents, e.g. chloroform and 3M aqueous NaClO$_4$, should depend only on the two solvents, provided that both values were obtained using the same counting assembly. We may therefore write

$$(N_i - N)Vk_w^{-1} = N_o V_o k_o^{-1} \qquad (9.16)$$

where N_i and N are the initial and final counts of the aqueous solution (of volume V) and N_o is count for the organic phase (of volume V_o). The difference $(N_i - N)$ should therefore be proportional to $N_o V_o V^{-1}$. It is advisable to obtain values of N_i, N and N_o, and to plot $(N_i - N)$ against $N_o V_o V^{-1}$ in order both to discard these points which do not fall on the line formed by the majority and to obtain the value of $k_w k_o^{-1}$ as the gradient. The value of the distribution ratio is then obtained as

$$q = \frac{(N_i - N)}{N} = \frac{N_o}{N} \frac{k_w}{k_o} \qquad (9.17)$$

Alternatively, the radioactivity may be transferred from each phase to a common medium which is used for all assays. For example, a carrier and a precipitant may be added. The precipitate is separated and each sample redissolved in the same volume of the same solvent (e.g. dilute acid of constant concentration). The distribution ratio is then simple the ratio of counting rates of samples obtained from the phases (e.g. equation (9.17) with $k_w/k_o = 1$).

Solvent extraction has been used, as described above, to study actinide complexes of a number of ligands such as acetylacetonate and 1-nitroso-2-naphtholate. Chloroform or isobutyl methyl ketone are suitable organic solvents. The radiochemical determination of distribution ratio is also an invaluable technique for studying solutions which contain extremely low concentrations of very insoluble complexes, such as 8-hydroxyquinolates, or hydroxides (see Fig. 9.4).

Partition of an uncharged complex MX$_c$ of an auxiliary ligand, X, between two liquids has been used to study complexes ML$_n$ which are formed only in the aqueous phase (see Fig. 9.5). This method is analogous to the use of an auxiliary ligand in the solubility method (see sect. 9.1). If the two sets of complexes MX and ML$_n$ exist in the aqueous layer, the distribution ratio is given by

$$q = \frac{(M_t)_0}{M_t} = \frac{[\mathrm{MX}_c]_0}{[\mathrm{M}] + \sum_1^N [\mathrm{MX}_x] + \sum_1^N [\mathrm{ML}_n]} = \frac{P_c[\mathrm{MX}_c]}{[\mathrm{M}] + \sum_1^N [\mathrm{MX}_x] + \sum_1^N [\mathrm{ML}_n]} \qquad (9.18)$$

or

$$P_c \beta_c [\mathrm{X}]^c q^{-1} = \sum_0^N \beta_n [\mathrm{L}]^n + \sum_1^N \beta_x [\mathrm{X}]^x \qquad (9.19)$$

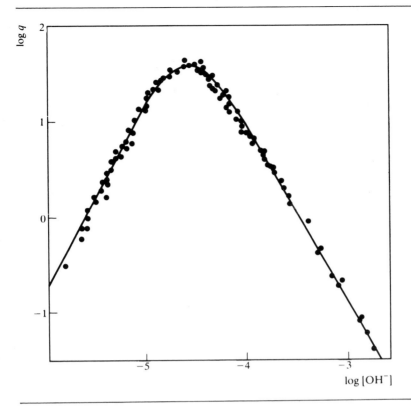

Fig. 9.4. Log q as a function of log [OH$^-$] for the extraction of Zn(OH)$_2$ from aqueous 3M NaClO$_4$ ($M_t \sim 5 \times 10^{-5}$M) into a solution of 5 per cent Amberlite CA1 in benzene. (After T. Sekine, *Acta Chem. Scand.*, **19**, 1526 (1965).)

Measurement of q as a function of [X] and [L] therefore leads to the function $\sum \beta_n [\mathrm{L}]^n$, hence to the stability constants β_n, provided that the parameters P_c and β_x for the complexes MX$_x$ have been measured.

Luckily, equation (9.19) may usually be considered simplified. Since the metal ion is usually present at only tracer concentration, X_t, $L_t \gg M_t$ so that [X] and [L] are independent of the extent of complex formation with M (though if X or L is basic, its free concentration will, of course, depend on [H] and on the appropriate protonation constant(s)). Moreover, the auxiliary ligand can sometimes be chosen so that partition takes place under conditions where no MX$_x$ complexes exist in the aqueous phase. Then the last term in equation (9.19) can be ignored, and when extraction is carried out at constant values of X_t and [H], but at varying L_t, we have

$$\left(\frac{q_{L_t=0}}{q} \right)_{X_t,[\mathrm{H}]} = \sum_{0}^{N} \beta_n [\mathrm{L}]^n \tag{9.20}$$

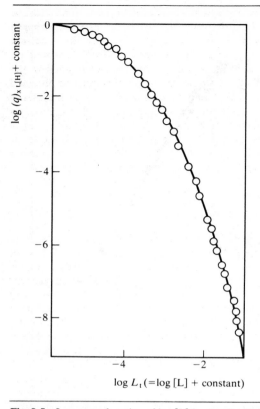

Fig. 9.5   Log q as a function of log [L] for the distribution of the Hf(IV) TTA complex between xylene and aqueous hydrofluoric acid, HL. (From B. Norén, *Acta Chem. Scand.*, **21**, 2435 (1967).)

from which the values of β_n may be calculated, without introducing values of $P_c\beta_c$ or β_x. Thenoyltrifluoroacetylacetone (TTA, I) is often used as an auxiliary extracting ligand in studies of complexes of non-basic inorganic anions, as its metal complexes are extracted into benzene from aqueous acid in which

$CO-CH_2-CO-CF_3$

(I) 2−thenoyltrifluouracetone ('TTA')

$H_t = [H]$ and $\sum_1 [MX_x] \sim 0$. Metal ion hydrolysis can also be studied with the help of TTA; the formation of hydroxo complexes suppresses extraction into benzene. The simple expression (9.20) cannot be used, since the necessary variation in [OH] leads to charges in [X] and hence in $[MX_x]$. Moreover, since the stability constants of MX_x complexes cannot be obtained in the absence of the hydroxo complexes, both sets of parameters must be obtained by successive approximations. This might seem to be an unnecessarily complicated method for studying hydrolysis. But, like other applications of liquid–liquid partition, it needs only tracer concentrations of metal ion and so has the enormous advantage that it can be used to obtain mononuclear hydrolysis constants for metal ions which form predominantly polynuclear hydroxo complexes at the concentrations required by other techniques.

9.3   Ion exchange

Just as uncharged species can partition between two immiscible liquids, ionic ones can partition between a solution and an ion exchanger. We shall illustrate the ion exchange method by discussing resin (rather than liquid) exchangers which consists of a negatively charged matrix which adsorbs cations. Suppose that all the sites on such a cation exchange resin are initially occupied by sodium ions. If this resin is equilibrated with a dilute solution of iron(III) chloride in the presence of a large excess of sodium perchlorate the various iron(III) cations Fe^{3+}, $FeCl^{2+}$ and $FeCl_2^+$ will be distributed between the solution and the resin (which will liberate the appropriate number of sodium ions to maintain its electrical neutrality). The distribution ratio of iron(III) between solution and resin is given by $q = (Fe_t)_R/[Fe]$ where $_R$ denotes the resin phase, in which concentrations may conveniently be expressed as moles of cationic species per gram of (usually air-dried) resin. For the iron(III) chloride system,

$$(Fe_t)_R = [Fe^{3+}]_R + [FeCl^{2+}]_R + [FeCl_2^+]_R$$
$$= P_0[Fe^{3+}] + P_1[FeCl^{2+}] + P_2[FeCl_2^+] \tag{(9.21)}$$

where $P_n = [FeCl_n]_R/[FeCl_n]$ is the partition coefficient of the nth complex. So, for the iron(III) chloride system

$$q = \frac{P_0[Fe^{3+}] + P_1[FeCl^{2+}] + P_2[FeCl_2^+]}{[Fe^{3+}] + [FeCl^{2+}] + [FeCl_2^+] + [FeCl_3] + \cdots} \tag{9.22}$$

and in general

$$q = \frac{\sum P_n\beta_n[L]^n}{\sum \beta_n[L]^n} \tag{9.23}$$

where values of $P_n > 0$ only for cations. It would be agreeable to be able to treat q as a polynomial in [L] only and so obtain values of β_n. Unfortunately, however, the activity coefficients in the resin phase, and hence the values of

P_n, are very sensitive to changes in the cationic content of the resin (see Fig. 9.6). And if the partition coefficients depend on $(M_t)_R$, then q is not strictly a function of [L] only.

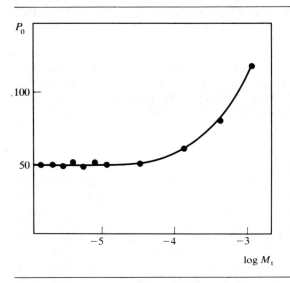

Fig. 9.6   The concentration-dependence of the partition coefficient P_0 for the distribution of uncomplexed Hf(IV) between aqueous solution of hafnium concentration M_t and Dowex 50 W resin. (From B. Norén, *Acta Chem. Scand.*, **21**, 2449 (1967).)

The distribution ratio for ion exchange, like that for solvent extraction, is best determined radiometrically using tracer concentrations of metal ion. But the values of P_c vary with $(M_t)_R$ even at these low concentrations. We may attempt to keep the values of P_n more nearly constant by measuring q at a number of different initial metal ion concentrations for each value of [L] to give a series of plots $q((M_t)_R)_{[L]}$ which are interpolated to give one function $q([L])_{(M_t)_R}$ which refers to the same 'load' $(M_t)_R$ on the resin. But even this time-consuming procedure does not guarantee constancy of partition coefficients because resins at constant load may differ considerably in composition according to the extent to which the adsorbed cations are complexed with ligand, and hence vary in charge.

Even if partition coefficients were true (i.e. constant) parameters, there are other disadvantages to the method. One drawback is the number of parameters needed to describe the data. If the free metal ion is the sole (non-background) cation, only one parameter is involved in addition to the stability constants; but for every cationic species formed, one more parameter is involved. And any increase in the number of parameters naturally decreases the precision with which any one of them can be determined.

Another drawback to the ion exchange method is the difficulty of obtaining separate samples of the two phases. Although the solvent can be freed from resin, the resin cannot be obtained from solvent since any attempt to wash it disturbs the equilibrium. The concentration of adsorbed metal ion must therefore be obtained from the difference $[M_t]_i - M_t$ between the initial and equilibrium concentrations in the aqueous phase as

$$(M_t)_R = ([M_t]_i - M_t)\frac{V}{m} \qquad (9.24)$$

where m is the mass of resin equilibrated with V l of solution. Since there can be no check on the mass balance of M in the ion exchange procedure, the value of q cannot be determined as precisely as it can for solvent extraction.

So the cation exchange method is more time-consuming than other distribution techniques, and yields less precise results which can only be interpreted by introducing more parameters (which cannot even be held constant). The method seems to have little to recommend it; but, in skilled and painstaking hands, it has nonetheless yielded some surprisingly good results.

9.4   Summary of distribution methods

Solvent extraction is probably the most useful of the three methods discussed. The value of M_t may be varied freely, down to tracer concentration if a suitable radioactive isotope is available. The distribution ratio can be measured conveniently, with moderate precision, and checked for conservation of total metal. Only one parameter, in addition to the stability constants, is needed; and since this partition coefficient refers to an uncharged species, often at low concentrations, its value is normally unaffected by such concentration changes as do occur. The method is fairly flexible and is particularly useful for studies of complexes which are present at very low concentration, e.g. actinide complexes, mononuclear hydrolysis products of metal ions, and complexes which are very sparingly soluble in water.

The solubility method is probably more precise, but more tedious, than solvent extraction, and the total metal ion concentration can be varied much less widely. The presence of the second phase again introduces only one additional parameter, and, provided that the solid phase is of constant composition and structure, activity coefficients in the solid are unity by definition. In the solubility method, as in solvent extraction, an auxiliary ligand can be introduced in order to study complexes which do not themselves take part in any distribution equilibria. However, although the method has no overwhelming disadvantage, it lacks any positive advantage.

The cation exchange method is, experimentally, both inconvenient and imprecise. Interpretation of measurements is also troublesome, since activity coefficients in the resin may vary, and a partition parameter is needed not only for the free metal ion, but also for each cationic complex. The method therefore has little to recommend it, except as a collector's item.

Several other types of distribution may be applied to the study of equilibria, e.g. measurement of the solubility of a salt NL_c of an auxiliary metal ion N, solvent extraction of an uncharged ligand (e.g. pyridine), partition chromatography, and the use of an anion exchange resin to study either partition of anionic complexes or an anionic ligand. But these methods have been used only seldom and will not be discussed further. The method has also been used to study the distribution of a metal ion between a fused salt medium and an organic solution of a complexing ligand.

Chapter 10

Experimental methods: III. Properties of the solution

In this chapter, we shall discuss those properties of the solution which can be exploited to give information about the formation of complexes. We shall limit the discussion to homogeneous systems, in which there is no second phase, nor even an electrode.

If a student is asked, out of the blue, how one might investigate the formation of complexes between the copper(II) ions and a simple ligand such as chloride, or ammonia, the most likely suggestion is 'spectrophotometry'. The popularity of this response is not surprising. We tend to be impressed with what we ourselves have seen and the dramatic deepening of colour which takes place when ammonia is added to copper sulphate solution may well have been our first encounter with complex formation. Of course, other less obvious changes in the properties of a solution may occur as the composition is changed and these, too, may be monitored. We shall attempt to identify some of these properties and to assess their convenience and reliability compared with those methods which we have already discussed. And, since it is unwise to belittle the obvious, we shall look first at the way in which the colour of a solution may be changed by complex formation.

10.1 Optical absorption

When a ligand displaces a solvent molecule from the coordination sphere of a metal ion, the energy levels of both the ligand and of the solvated metal ion will be altered. If any one of the species involved absorbs visible light, complex formation will therefore be associated with some change in colour; and exactly analogous, though invisible, changes occur if any of the species absorbs ultraviolet radiation. Since many transition metal ions are coloured and many organic ligands absorb ultraviolet radiation, measurement of the optical absorption of a solution would seem a promising technique for monitoring complex formation.

A commercial spectrophotometer normally produces an absorption spectrum in the form of a plot of optical absorbency, A_s, against wavelength, where

$$A_s = \log I_0/I \qquad (10.1)$$

depends on the ratio of the intensities I_0 and I of incident and transmitted radiation of each particular wavelength λ. The wavelength λ_{max} at which a peak occurs in the absorption spectrum indicates the energy of the transition, whereas the height of the peak is determined by the number of species which are undergoing transition. So the intensity of absorption depends on the length of the sample through which the radiation has to pass, the concentration of the species S which is capable of being excited by radiation of wavelength λ_{max}, and the proportion of the total number of S ions or molecules which in fact undergo the transition. When a single species S is absorbing at wavelength λ_{max}, the absorbency of the solution is given by the Beer–Lambert law

$$A_s = l\varepsilon_S[S] \tag{10.2}$$

where l is the length of optical path and ε_S is the extinction coefficient of S. When several absorbing species are present, the total absorbency is merely the sum of the absorbencies of the separate species, so that

$$A_s = l \sum_S \varepsilon_S[S] \tag{10.3}$$

The extinction coefficients ε_S are true parameters in that they are constant for a particular species at a particular wavelength and are (very nearly) independent of the ionic environment. They describe the probability of a transition. Allowed transitions, as in charge transfer complexes such as $FeSCN^{2+}$, give rise to very high values of ε_S ($\sim 10^3 \, mol^{-1} \, cm^{-1}$) while lower values ($< 10^2 \, mol^{-1} \, cm^{-1}$) are obtained for 'forbidden' (i.e. less probable) transitions, such as those which give rise to the characteristic colours of Cu^{2+} aq, $Cu(NH_3)_4^{2+}$ and Co^{2+} aq. The use of [S] rather than {S} in the Beer–Lambert equation is a rigorous description of the fact that the absorbency of a species is indeed proportional to its concentration (where the concentration is a measure of a number of solute species in a given quantity of solvent rather than a euphemism for activity, linked to an optimistic assumption that activity coefficients can be kept roughly constant). At first sight, the absence of activity coefficients in the Beer–Lambert equation might seem to give spectrophotometry a great advantage over the activity-based 'thermodynamic' methods discussed in the previous two chapters. But the gain is illusory. Although it is comforting to know that extinction coefficients are little influenced by their environment, we have seen that stepwise equilibria can seldom be handled satisfactorily unless the stoichiometric stability constants are held constant; and so a bulk electrolyte must in any case be used. Spectrophotometry does, however, have one very real advantage over most other techniques used in equilibrium chemistry; ease of measurement. Given standard commercial equipment and satisfactory stock solutions, only routine manual skill is necessary to obtain measurements of adequate precision extremely rapidly. The main problem is the analysis of the results.

When a solution containing a metal ion M, a ligand L and a series of mononuclear complexes ML_n is irradiated with monochromatic light in an optical cell of length l, its optical absorbency is given by

$$A_s = l\left\{ \varepsilon_L[L] + \sum_{n=0}^{N} \varepsilon_n[ML_n] \right\} \tag{10.4}$$

where ε_n is the extinction coefficient of the nth complex at the particular wavelength used. It is often convenient to discuss absorption of radiation by the complexes in terms of an 'observed' extinction coefficient, defined by

$$\mathscr{E} = \frac{\sum \varepsilon_n [\mathrm{ML}_n]}{\sum [\mathrm{ML}_n]} \tag{10.5}$$

and related to the measured value of the optical absorbency by

$$\mathscr{E} = \frac{A_s l^{-1} - \varepsilon_L [\mathrm{L}]}{M_t} \tag{10.6}$$

Measurements are often carried out at wavelengths where the metal ion, and its complexes, absorb, but the ligand does not. Equation (10.6) then reduces to

$$\mathscr{E} = \frac{A_s}{l M_t} \tag{10.7}$$

Since M_t is known from the composition of the solution, and l from the manufacturer's specifications, the value of $\mathscr{E}$ may readily be obtained. We may now follow the familiar procedure of replacing the variables $[\mathrm{ML}_n]$ by the products $\beta_n [\mathrm{M}][\mathrm{L}]^n$ and since the system contains only mononuclear complexes, we may cancel out $[\mathrm{M}]$ to give

$$\mathscr{E} = \frac{\sum \varepsilon_n \beta_n [\mathrm{L}]^n}{\sum \beta_n [\mathrm{L}]^n} \tag{10.8}$$

Thus in place of equation (10.5), which contains $(N+1)$ concentration variables, we have a relationship between $\mathscr{E}$ and the single variable $[\mathrm{L}]$. But we have paid the usual price for this type of simplification: one extra parameter has appeared for every variable eliminated. Equation (10.8) indicates that values of ε_n and β_n can in principle be obtained from at least $(2N+1)$ pairs of values of $\mathscr{E}$ and $[\mathrm{L}]$. We have seen that values of $\mathscr{E}$ may readily be obtained from values of the optical absorbency of solutions in which M_t is known, but evaluation of $[\mathrm{L}]$ poses more of a problem. If one or more of the species absorbs very strongly, it may be possible to use such dilute solutions of M that a negligible proportion of the ligand is bound to it; and if $\bar{n} M_t$ is indeed negligible compared with L_t, we may put $[\mathrm{L}] = L_t$ for a non-basic ligand and $[\mathrm{L}] = L_t / (\sum_0^J \beta_j [\mathrm{H}]^j)$ for a basic one. In the latter type of system, the value of $[\mathrm{H}]$ must be measured, and the protonation constants of the ligand known.

The plethora of parameters in equation (10.8) is, of course, a great disadvantage, because it makes the calculation more difficult and decreases the precision of the results obtained. And although an electronic computer may largely take care of the first factor, it is powerless to improve on the second. The greater the number of parameters which can be adjusted to describe the data, the less can we rely on the value of any one of them, regardless of the type of machine by which we calculated them.

If $\bar{n} M_t$ is not negligible compared with L_t, we must nonetheless proceed as if it were and obtain preliminary value of β_n (and of ε_n). We use these approximate stability constants together with $L_t(\sim[L])$, to obtain rough values of $\bar{n}$ and hence of [L] by means of equations (7.20) and (7.16). The values of β_n and, incidentally, of ε_n, are refined by successive approximations until they converge: a daunting prospect for which a computer is virtually essential.

So, despite the ease with which we can measure absorbency, and the consoling constancy of the extinction coefficients, spectrophotometry does not seem a very promising method for studying stepwise equilibria. It is, however, sometimes very useful for determining secondary concentration variables, in two quite different situations:

(i) *Determination of* α_c If a single species absorbs at a particular wavelength, the absorbency is proportional to the concentration of that species. The value of α_c may then be obtained provided that the extinction coefficient of ML_c can be measured. Since the spectrum of a hydrated metal ion may be grossly changed by complex formation, it may not be too difficult to find a wavelength where only the free metal ion absorbs. For example, the colour of an aqueous solution of iron(III) changes from yellow to purple when phenol is added. In systems of highly polydentate ligands, where only one complex is formed, it is again quite likely that the complex will absorb at some wavelength where the free metal ion does not. The value of α_1 may be obtained from the absorbency of solutions which contain such high concentrations of ligand that effectively all the metal is in the form of the complex, and then used to calculate the value of α_1 in partially complexed solutions. If both M and ML absorb, but to a markedly different extent, measurements of absorbency may be used to obtain the value of β_1 in an exactly analogous way to that described for the spectrophotometric determination of the protonation constant of a monobasic acid.

(ii) *Determination of* $\bar{n}$ We have seen from equation (10.8) that $\mathscr{E}$ is a function of the single variable [L]. So, if two solutions which contain different total concentrations of both metal ion and ligand are found to have the same value of $\mathscr{E}$, they must both contain the same concentration of uncomplexed ligand. The values of $\bar{n}$, and of α_c for each complex, must also be the same in the two solutions, since these secondary concentration variables are also functions only of [L]. Any two such solutions, in which [L] is the same but M_t and L_t differ, are said to be 'corresponding'. Now from equation (7.16), $\bar{n}M_t = L_t - [L]$ so that, for a set of corresponding solutions in which, by definition, [L] and $\bar{n}$ have fixed values, a plot of L_t against M_t is a straight line of gradient $\bar{n}$ and of intercept [L]. Spectrophotometry is a very convenient method for measuring $\bar{n}$ by the corresponding solutions method (see Fig. 10.1). First, A_s is measured as a function of L_t for a series of solutions in which M_t is held constant; and the procedure is repeated for a number of other sets of solutions each with a different, but constant, value of M_t. The family of curves $A_s(L_t)_{M_t}$ which is so obtained is then interpolated at a number of constant values of $\mathscr{E}$ and [L] to give pairs of values of L_t and M_t for sets of corresponding solutions, each set referring to a different value of $\mathscr{E}$. By using optical cells of different thickness, and techniques for studying samples of high absorbency, quite wide variations in M_t and L_t may be achieved. For each set, values of $\bar{n}$ and [L] are obtained from the linear plot

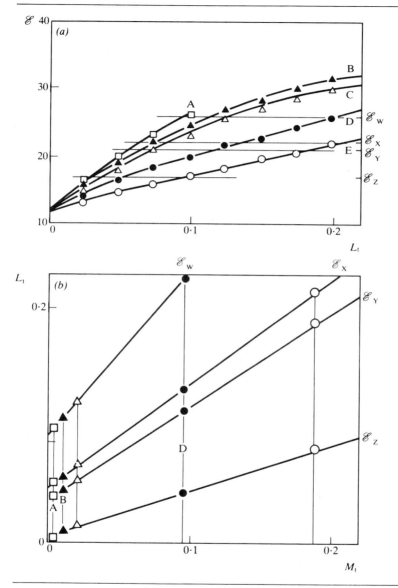

Fig. 10.1 The spectroscopic determination of $\bar{n}$ and [L] for VO^{2+}–SCN^- by the method of corresponding solutions. (*a*) Observed extinction coefficient $\mathscr{E}$ at 700 nm as a function of L_t for solutions of different total concentration M_t of metal ion ranging from $4\cdot7\times10^{-3}$M (solution A) to $1\cdot878\times10^{-1}$M (solution E). (*b*) Plots of L_t against M_t for sets of corresponding solutions obtained by interpolating the plots in Fig. 10.1(*a*) at constant values $\mathscr{E}_w, \ldots, \mathscr{E}_z$ of $\mathscr{E}$. The lines $L_t(M_t)_\mathscr{E}$ are of slope $\bar{n}$ and of intercept [L]. (After F. J. C. Rossotti and H. Rossotti, *The Determination of Stability Constants*, McGraw-Hill, New York (1961).)

of M_t against L_t. The method can be used for systems in which several complexes are formed, provided that all are mononuclear. But the interpolation procedure requires a large number of measurements and results in loss of precision. In careful hands, however, the method gives useful results for systems which cannot be studied potentiometrically.

The most frequent use of spectrophotometric data in the study of equilibria has, regrettably, been in a procedure usually known as the method of continuous variations, or Job's method. The absorbency (or any other property which changes in a fashion which is proportional to the concentration of the complex ML_n) is measured for a series of solutions in which $(M_t + L_t)$ is held constant whilst varying the ratio of M_t/L_t. If a single complex ML_n is present, the absorbency reaches a maximum value when $L_t/M_t = n$. The stability constant of ML_n is calculated from the absorbency at this maximum, as follows. At high ratios L_t/M_t, it is assumed that all the metal is complexed so that $[ML_n] = M_t$ and $A_s l^{-1} = \varepsilon_n M_t$. Similarly, at low values of L_t/M_t, all the ligand is assumed to be bound to metal, so that $[ML_n] = L_t/n$ and $A_s l^{-1} = \varepsilon_n L_t/n$. The tangents to the Job plot at the limiting points $M_t \to 0$ and $L_t \to 0$ therefore intersect at an ordinate corresponding to the absorbency of a complex of concentration $[ML_n] = M_t = L_t/n$. Since the observed maximal absorbency gives the actual concentration of the complex at the same total concentrations of metal and ligand, the equilibrium constant may readily be calculated.

In practice, of course, the method is restricted to systems which contain only the species M, L and ML since solutions which contain appreciable concentrations of, say, M, L and ML_3 also contain appreciable concentrations of intermediate species. Attempts which have frequently been made to extend Job's method to more complicated systems have proved unsatisfactory because it is so difficult to draw unambiguous conclusions about maxima in any position other than where $M_t = L_t$. Since those systems in which only one complex is formed may be studied fairly easily by the spectrophotometric methods described above, Job's method has little to recommend it and it seems odd that is has survived to the present day.

Most spectrophotometric studies of metal complexes make use of changes caused by the ligand in the visible absorption spectrum of the metal ion. Although the formation of the first complex may give rise to an appreciable change in colour, addition of further ligands would be expected to have much less effect since there is often considerable overlap between the absorption bands of the various complexes. It is therefore difficult to interpret spectrophotometric measurements on solutions which contain two or more complexes, particularly if no electronic computer is available.

The addition of an anionic ligand to an aqueous metal ion sometimes alters the ultraviolet spectrum without affecting the absorption of visible light. Such changes have been thought to indicate the presence of an 'outer-sphere' complex in which the aquo cation forms an ion pair with the anion without losing any water from its inner hydration sphere. But since this interpretation is open to criticism, spectrophotometric values of equilibrium constants of reactions involving outer sphere complexes should be viewed with caution.

Other types of spectroscopy have been far less frequently used for measuring equilibrium constants and have been mainly restricted to systems in which ML is the sole complex formed. Since water absorbs strongly in the near infra-red, spectrophotometry in this region has been used only for non-aqueous systems and for those few aqueous ones which contain species, such as CN^-, SCN^- and VO^{2+}, which have strong, narrow absorption bands at frequencies well removed from the main peaks in the spectrum of water. Raman spectroscopy, even with a high-intensity laser source, requires such concentrated solutions that control of activity coefficients is difficult. High-resolution nuclear magnetic resonance is a potentially powerful method, since the concentrations of a number of coexistent species may be measured, but it has not yet been widely used.

Formally similar methods

A number of other properties $\mathscr{P}$ may be analysed in the same way as we have described for spectrophotometric data, provided that they depend on concentrations, and intensive factors, i_S, according to the relationship

$$\mathscr{P} = \sum_S i_S[S] \tag{10.9}$$

Equation (10.3) is, of course, a special case of (10.9) with $\mathscr{P} = A_s$ and $i_S = \varepsilon_S$.

The method has been used to determine stability constants of complexes with optically active ligands. Here $\mathscr{P}$ is the observed optical rotation of the solution and i_S are the molar rotations of the various optically active species. The heat of reaction between A and B may be similarly interpreted in terms of the heats of formation of the various complexes.

10.2 Colligative properties

These properties include the osmotic pressure, the elevation of boiling point and the lowering of the vapour pressure and of the freezing point. All depend on the effect of the total concentration of the solute species on the activity of the solvent and are independent of the nature of the solutes. Only the last property, cryoscopy, has been used in the study of the metal complexes. Unless activity coefficients are held constant, interpretation of the measurements is extremely difficult. Meaningful results are best obtained by using as ionic medium a saturated solution of an inert electrolyte and measuring the effect of added metal ions and ligand on the eutectic temperature.

If the bulk electrolyte is sufficiently concentrated to ensure that the activity coefficients are constant, the lowering ΔT of the eutectic or transition temperature is given by

$$\Delta T k^{-1} = S = [L] + [M] + [ML] + [ML_2] + \cdots \tag{10.10}$$

where k is the cryoscopic constant for that particular medium and (approximate) temperature and S is the sum of the concentrations of all species other than the ions of the bulk electrolyte. Since

$$M_t = \sum_0^N [ML_n]$$

and

$$L_t = [L] + \sum_0^N n[ML_n]$$

the total number of solute species is given by

$$S = L_t + M_t - \sum_0^N n[ML_n] = L_t + M_t(1 - \bar{n}) \qquad (10.11)$$

whence values of

$$\bar{n} = \frac{L_t + M_t - S}{M_t} \qquad (10.12)$$

and hence of

$$[L] = L_t - \bar{n}M_t \qquad (10.13)$$

may be obtained. The stability constants may be calculated from the function $\bar{n}([L])$ in the usual way (see Ch. 12). As in other studies of complex formation in a bulk electrolyte, it is assumed that only counter-ions added to the solution together with M and L are those which also comprise the ionic medium.

Cryoscopy is a very inflexible method for studying equilibria and has little to recommend it. For a given background electrolyte, we are restricted to one concentration and to one temperature; and solutions containing different concentrations of solute are not strictly isothermal. The method cannot be used if the ligand can become protonated or if the metal ions are hydrolysed, and is best considered only as a collector's item.

10.3   Electrical conductivity

One might suppose that measurement of electrical conductivity would be a useful way of monitoring complex formation between a metal ion and an anionic ligand. Indeed, the very concept of electrolytic dissociation is based mainly on conductimetric studies.

The equivalent conductivity Λ of a solution depends on the value Λ_s for the pure solvent and on the concentrations, charges and mobilities u_L and u_n of the ligand L and complexes ML_n. Thus

$$\frac{\Lambda - \Lambda_s}{10^3} = \frac{u_L[L]}{L_t} + \frac{(z_M - nz_L)u_n\beta_n[L]}{z_M \sum \beta_n[L]} \qquad (10.14)$$

This expression is similar to equation (10.8), which describes the dependence of optical absorbency on free ligand concentration; but it serves to emphasize the many disadvantages of conductivity as a method for studying complex formation. If measurable changes in conductivity are to be obtained, the value of Λ_s should be negligible compared with Λ. The use of a constant ionic medium is therefore out of the question, and so any change in L_t or M_t will lead to variations in activity coefficients and hence in the values of β_n. Nor are the ionic mobilities constant. On the contrary, they are extremely sensitive to ionic interaction, and hence to the equilibrium composition of the solution, in a way which is extremely hard to predict, even in the absence of complex formation. The coefficients of [L] in equation (10.14) contain $(2N+1)$ unknown quantities which, unlike their analogues in the spectrophotometric relationship (10.8), are not even constants. Quantitative interpretation of conductivities is therefore extremely difficult. Reliable values of $_a\beta_1$ have been obtained from very precise work on 1:1 electrolytes, but the method is not recommended for any but the very simplest systems.

10.4 Summary

Of the three main types of method discussed here, spectrophotometry is, despite its disadvantages, by far the best. There is great flexibility in the choice of medium and temperature. Activity coefficients may therefore be controlled and the extinction coefficients are effectively constant. Since measurements may be made over a wide range of wavelengths, it may be possible to eliminate the absorption of some species (e.g. the free ligand) or to measure the concentration of one species only. In the general case, however, there are $(2N+2)$ parameters.

Cryoscopy is much more limited since measurements are restricted to only a few media, and to a single, fixed temperature for each. All species contribute to the measurements, but as no intensive factors are introduced (except for the cryoscopic constant of the solvent) there are only N disposable parameters.

Conductivity is the least satisfactory of the three techniques. All charged species contribute, so that there are $(2N+1)$ coefficients of terms in [L]; but an ionic medium cannot be used, and none of these coefficients remain constant throughout the measurements.

Any other property of a solution which changes on complex formation can, in principle, be used to study equilibria in solution. Many different techniques have been used, for example measurements of magnetic susceptibility, dielectric constant and rate of reaction. But since note of them has been widely used, they will not be discussed further.

Chapter 11

Complications

So far, we have discussed only very simple systems in which all metal complexes ML_n are mononuclear with respect to the metal ion, all acids H_jL are mononuclear with respect to ligand L, and no other complexes are formed. Happily, these conditions are fulfilled for a large number of combinations of metal ion and ligand under just those conditions (e.g. $10^{-1}M \gtrsim M_t \gtrsim 10^{-3}M$ and $1 \lesssim pH \lesssim 5$) which are most conveniently used for studying equilibria. But there is nonetheless scope for the formation of more complicated species. For example, the metal ion may be hydrolysed, and many metal hydroxo complexes, e.g. $Fe_2(OH)_2^{4+}$, are polynuclear. The complexes ML_n may themselves be hydrolysed, e.g. $Hg(OH)I$, or protonated, e.g. $HFeCl_4$. The metal ion must then be considered as binding two different ligands, L and H (or OH), within the same complex. Similar ternary species may be formed when two different ligands not derived from the solvent are added to a solution of metal ions. Iron(III), for example, forms mixed halide and pseudo-halide complexes, such as $Fe(SCN)Cl^+$. The acids derived from L may dimerise, as in acetate solutions where the species H_2L_2 and HL_2^- appear to be formed. The binding of metal ions to proteins is similar to the binding of protons in that many ions may be associated with a single protein, and so it is convenient to treat the protein as 'central group' and the cations as 'complexing agents'. Several uncharged species, such as phenol, undergo oligomerisation, or self-association, in solution, often by intermolecular hydrogen bonding; and some carboxylate ions associate to form micelles, such as X_{15}^{15-}, where X^- is the 3-methyl butyrate ion.

The possibility that one or more of these more complicated species may be formed makes it essential to test for their presence before we can properly apply any treatment which assumes the equilibria involve only the complexes ML_n and H_jL. If polynuclear or mixed complexes are indeed found to be present, we may proceed in one of two ways, depending on how we view these species. Those who are primarily interested in the simple mononuclear complexes probably consider more complicated species merely as an interference. It is then best to try to eliminate the nuisance value of mixed or polynuclear species by working under conditions where they are formed only to a negligible extent. On the other hand, if study of the more complicated species is the main object of an investigation, measurements should obviously be made under conditions (and preferably under a wide range of conditions) where such complexes are predominant, or at least exist at appreciable concentrations.

Not surprisingly, the relatively simple equations which we have used so far become much more complicated if they are adapted for use with polynuclear or mixed complexes, even though we shall use essentially the same procedure, viz:

1. Define an equilibrium constant for the formation of each species.
2. Express the (known) total concentrations of the components M, L, H, etc. in terms of the equilibrium concentrations of all species present.
3. Replace the equilibrium concentrations of the complexes by the product of the appropriate equilibrium concentrations and the equilibrium constant.
4. Measure the equilibrium concentration of one (or more) species as a function of the various total concentrations.
5. Eliminate the free concentrations of the other components to give an expression in one (measurable) equilibrium concentration, and as many of the total concentrations as are needed.
6. Substitute the experimental values of these concentration variables into the theoretical relationship, and solve the equations so obtained for the required values of the equilibrium constant.

Steps 1 to 3 present little difficulty. It is again assumed that activity coefficients can be controlled, so that any complex $A_a B_b C_c \dots$ can be particularised by its overall stoichiometric stability constant

$$\beta_{a,b,c\dots} = \frac{[A_a B_b C_c \dots]}{[A]^a [B]^b [C]^c \dots} \tag{11.1}$$

The total concentration of, for example, A is given by

$$A_t = \sum_{a=0} \sum_{b=0} \sum_{c=0} \cdots a[A_a B_b C_c \dots]$$

$$= \sum_{a=0} \sum_{b=0} \sum_{c=0} \cdots a\beta_{a,b,c\dots}[A]^a [B]^b [C]^c \dots \tag{11.2}$$

Steps 5 and 6 can, however, be extremely troublesome. They are best performed by computer and are most likely to yield meaningful results if the experimental step, 4, has been carried out with maximal care.

We shall illustrate the use of equations (11.1) and (11.2) with a few examples.

11.1 Polynuclear complexes $M_m L_l$

Let us suppose, for simplicity, that [M] can be measured with a suitable electrode and that L is a halide ion so that its concentration, too, can be measured potentiometrically. The total concentrations are given by

$$M_t = \sum_{m=0} \sum_{l=0} m[M_m L_l]$$

$$= \sum_{m=0} \sum_{l=0} m\beta_{ml}[M]^m [L]^l \tag{11.3}$$

and

$$L_t = \sum_{m=0} \sum_{l=0} l\beta_{ml}[M]^m[L]^l \qquad (11.4)$$

Since we can measure both [M] and [L] we can obtain experimental values of α_0 and $\bar{n}$, which we shall define exactly as we did for mononuclear systems in Chapter 7, so that

$$\alpha_0 = \frac{[M]}{M_t} \qquad (7.21)$$

and

$$\bar{n} = \frac{(L_t - [L])}{M_t} \qquad (7.16)$$

Combination of equations (11.3) and (11.4) with (7.21) and (7.16) gives

$$\alpha_0 = \frac{1}{\displaystyle\sum_{m=0} \sum_{l=0} m\beta_{ml}[M]^{m-l}[L]^l} \qquad (11.5)$$

and

$$\bar{n} = \frac{\displaystyle\sum_{m=1} \sum_{l=0} l\beta_{ml}[M]^m[L]^l}{\displaystyle\sum_{m=0} \sum_{l=0} m\beta_{ml}[M]^m[L]^l} \qquad (11.6)$$

Thus, if polynuclear complexes are formed, the secondary variables α_0 and $\bar{n}$ are functions not only of [L], but also of [M]. Experimental plots of α_0 or $\bar{n}$ as functions log [L] obtained using different initial concentrations of metal ion will therefore not coincide (see Fig. 11.1). Conversely, if functions $\alpha_0([L])$ and $\bar{n}([L])$ are found to be independent of the total concentration of metal ion, we may infer that only mononuclear complexes are present in appreciable concentrations.

The reason for the familiar plea that studies of metal complex formation should be carried out using several metal ion concentrations is now clear: it enables us to ascertain whether or not polynuclear complexes are formed. If plots of $\bar{n}$, or α_0, against log [L] give widely separated curves for different values of M_t, we know that polynuclear complexes are present in appreciable concentrations. Sets of experimental values of [M], [L], M_t and L_t must then be substituted into equations (11.5) and (11.6), which are then solved for the parameters β_{ml}. A value of β_{ml} which is appreciably greater than zero indicates that the particular complex M_mL_n can be formed in significant concentrations, whereas a value of β_{ml} within the range $0 \pm x$ (where x is the expected experimental error in the equilibrium constant) provides no evidence for the existence of M_mL_l.

If the same plot of $\bar{n}$ against log [L] or of α_0 against log [L] is obtained using a number of different initial concentrations of metal ion, we may deduce that only mononuclear species are formed in appreciable concentrations *within the concentration range used*.

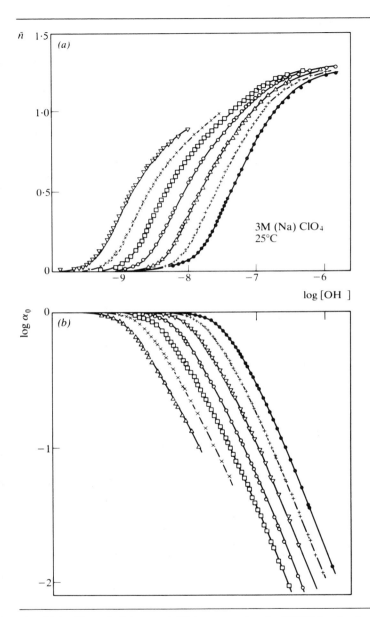

Fig. 11.1 Plots of: (*a*) $\bar{n}$; and (*b*) $\log \alpha_c$, against $\log[OH^-]$ for the hydrolysis of lead(II) showing dependence on M_t which ranges from 0·08M (for the left-hand solution) to 0·00125M. The main complexes $Pb_q(OH)_p^{(2q-p)-}$ are those for which q, $p = 1,1$; 4,4; 3,4; and 6,8. (After Å. Olin, *Acta Chem. Scand.*, **14**, 126 (1960).)

It has been assumed throughout this section that the ligand L does not bind protons and that its concentration can be measured directly. There are, of course, only few ligands for which this is true, since most species which combine with metal ions also do so with hydrogen ions. However, the concentration of hydroxyl ions may readily be measured potentiometrically provided that the ionic product of water has been determined under the same conditions. And hydroxo complexes of metal ions are notoriously polynuclear. The total hydroxyl ion concentration is given by

$$L_t = (OH)_t = -H_t = \sum_{q=1} \sum_{p=1} p[M_q(OH)_p] + [OH] - [H] \tag{11.7}$$

In the more general case where L is any basic ligand other than the hydroxyl ion, we need to know the total concentrations of both L_t and H_t, so that, if the acids H_jL are mononuclear, we have

$$L_t = \sum_{q=1} \sum_{p=1} p[M_qL_p] + \sum_{j=0} [H_jL] \tag{11.8}$$

and

$$H_t = \sum_{j=0} j[H_jL] + [H] - [OH] \tag{11.9}$$

It is, of course, by no means always possible to obtain values of both [M] and [L] experimentally. In such cases, one of the concentration variables is measured as a function of M_t, L_t (and if the ligand is basic, also of H_t). The other concentration variable may then, in principle, be obtained by one of a number of sophisticated computational or graphical methods, all of which demand a large number of experimental data of the highest possible precision.

Interpretation of data is easiest for systems in which a single polynuclear complex is formed, or for those which exhibit one of a number of set patterns of behaviour, which are outlined below. If measurements can be made over a wide range of concentrations it may be possible to simplify the process of interpretation by considering data from restricted concentration ranges. Thus we can sometimes study mononuclear complexes using solutions which are so dilute that polynuclear complexes are absent; and, conversely, obtain information about polynuclear complexes from regions of such high concentration that mononuclear complex formation is negligible. But we must then satisfy ourselves that measurements obtained at intermediate concentrations are compatible with the coexistence of both sets of complexes.

The following types of relatively simple behaviour may be encountered in polynuclear systems:

(i) *Formation of a unique complex*   If a single complex M_QL_P is formed, its formula and stability constant may be obtained fairly easily, particularly if both [M] and [L] can be measured. A number of solutions of hydrolysed metal ions have been found to contain a single polynuclear complex, and for some of these the free metal ion concentration, as well as $[OH^-] = K_w[H]^{-1}$, can be measured potentiometrically. From equations (11.3) and 11.4) we may then write

$$[M_Q(OH)_P] = \frac{M_t - [M]}{Q} \tag{11.10}$$

$$= \frac{[H] - [OH] + H_t}{P} \tag{11.11}$$

so that the empirical formula may be obtained from the ratio

$$\frac{P}{Q} = \frac{([H^+] - [OH] + H_t)}{(M_t - [M])} \tag{11.12}$$

Since the stability constant of the complex may be written as

$$\beta_{QP} = \frac{(M_t - [M])Q^{-1}}{([M][L]^{P/Q})^Q} \tag{11.13}$$

we may obtain the value of Q as the slope of the linear plot of $\log (M_t - [M])$ against $\log [M][L]^{P/Q}$. The value of $\log Q\beta_{QP}$ is obtained as the intercept. This method is well suited to the study of the hydrolysis of iron(III) under conditions where the metal is largely in the form either of the unhydrolysed aquo ion or of the dimeric hydroxo complex $Fe_2(OH)_2^{4+}$. The concentration of Fe^{3+} can be determined using an $[Fe^{3+}]$, $[Fe^{2+}]$ redox electrode, because iron(II) is not appreciably hydrolysed in solutions more acidic than pH ~ 7. The hydroxyl ion concentration can be obtained from measurements with a glass electrode.

Even when it is not possible to measure $[M]$, it is not too difficult to obtain information about the formation of a unique complex from the data M_t, L_t and $[L]$. We may combine equations (11.3) and (11.6) to give

$$M_QL_P = \frac{M_t - [M]}{Q} = \frac{\bar{n}M_t}{P} = \beta_{QP}[M]^Q[L]^P \tag{11.14}$$

elimination of $[M] = M_t(1 - Q\bar{n}/P)$ gives

$$\frac{\bar{n}}{(P - \bar{n}Q)^Q} = \beta_{QP}P^{1-Q}M_t^{Q-1}[L]^P \tag{11.15}$$

Thus

$$\left(\frac{\partial \log M_t}{\partial \log [L]}\right)_{\bar{n}} = \frac{P}{Q-1} \tag{11.16}$$

and

$$\left(\frac{\partial \log \bar{n}}{\partial \log [L]}\right)_{M_t} = P \tag{11.17}$$

Thus, provided that there is a generous supply of precise measurements, the values of P and Q may be obtained from the experimental functions $\bar{n}([L])_{M_t}$ and from the interpolated functions $M_t([L])_{\bar{n}}$. Once the values of P and Q

have been obtained, the logarithm of the left-hand side of equation (11.15) may be plotted against $((Q-1)\log M_t + P\log[L])$ to give a straight line of unit gradient and of intercept $(\log \beta_{QP} + (1-Q)\log P)$ (see Fig. 11.2).

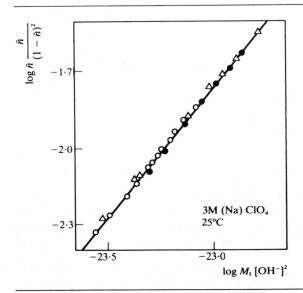

Fig. 11.2 The determination of β_{QP} for the unique complex $(VO)_2(OH)_2^{2+}$ by means of a linear plot, based on equation (11.15) using $P=2$ and $Q=2$ in the range $0.05M > M_t > 0.01M$. (After F. J. C. Rossotti and H. Rossotti, *Acta Chem. Scand.*, **9**, 1177 (1955).)

For systems which are studied in this way, and indeed for most work on polynuclear complexes, it is convenient to keep M_t constant during a set of experiments in which [H] or [L] is measured with varying values of H_t or L_t. In potentiometric titrations this may easily be achieved by using an initial concentration M_t of metal ions in the titration vessel and titrating with equal aliquots of two solutions:

(a) the solution used to vary H_t or L_t (containing strong acid, strong base, or ligand, but no metal M); and

(b) a solution containing metal ions at a concentration of $2M_t$. The solution will probably also contain acid, the concentration of which must, of course, be taken into account when calculating the value of H_t at each point in the titration.

(ii) *Homonuclear complexes*   If all metal ion species M_qL_p have the same degree of condensation (i.e. the same value, Q, of q) the system is homonuclear, and addition of ligands to the unchanging central group M_Q can then be treated in just the same way as the formation of mononuclear complexes of an uncondensed metal ion. The total, and free, concentrations of central group are M_t/Q and $[M_Q]$. Mercury(I), which exists as Hg_2^{2+} in the aquo ion and in all known complexes, is a familiar example.

(iii) *Cross-over points*   If an ion M exists in only two states of condensation (i.e. can form species with only two values of q) we can show that there will be one free ligand concentration $[L]^*$ at which the value $\bar{n}^*$ of $\bar{n}$ is independent of M_t. Plots of $\bar{n}$ against $\log [L]$ will therefore all intersect at a unique point, $\bar{n}^*$, $\log [L]^*$. Although it is not theoretically impossible that a cross-over point could be observed in a system with more than two degrees of condensation, by far the most likely cause of a cross-over point is the presence of two series of complexes $M_q L_p$ and $M_{q'} L_p$; and one of these series may well be mononuclear. In the simplest case where a unique polynuclear complex $M_Q L_P$ exists with mononuclear complexes, the value of $\bar{n}^*$ equals the ratio P/Q. Methods for identifying the complexes in more complicated systems which show cross-over points and for obtaining values of stability constants have been used to study proton–carboxylate equilibria and the formation of polyacids, such as borates (see Fig. 11.3).

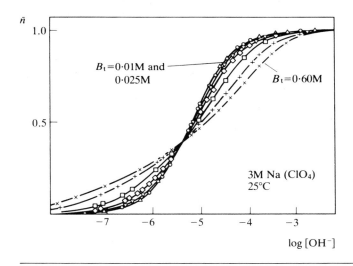

Fig. 11.3   A cross-over point in plots of $\bar{n}$ against $\log [OH^-]$ for the addition of OH^- ions to boric acid $B(OH)_3$. The main complexes are $B(OH)_4^-$, $B_3O_5(OH)_4^-$ and $B_3O_3(OH)_5^{2-}$. (From N. Ingri, C. Lagerström, M. Frydman and L. G. Sillén, *Acta Chem. Scand.*, **11**, 1034 (1957).)

(iv) *Parallel plots*   For some solutions of hydrolysed metal ions the families of curves $\bar{n}(\log [L])_{M_t}$ and $\alpha_0(\log [L])_{M_t}$ are found to be parallel (see Fig. 11.4). This behaviour is observed if a single polynuclear complex is present, but it may also arise from the formation of series of complexes. Interpretation is then much more difficult, and there may be uncertainty even in the exact formulae of the complexes, let alone in their stability constants.

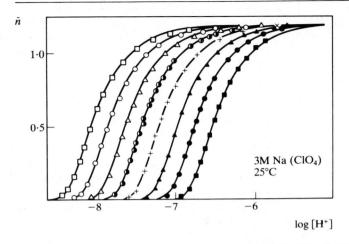

Fig. 11.4 Parallel curves $\bar{n}(\log[L])$ for the addition of protons to tungstate ions WO_4^{2-} for values of M_t ranging from 0·160M (for the left-hand solution) to 0·0012M. The data are compatible with the formation of a single complex $HW_6O_{21}^{5-}$. (After Y. Sasaki, *Acta Chem. Scand.*, **15**, 175 (1961).)

Sillén showed that systems which gave parallel curves could be treated as if complex formation took place by combination of the simple metal ion M with a series of identical composite ligands (M_aL_b) which are themselves complexes. It was moreover supposed that the equilibrium constants for the addition of successive composite groups were not independent; but could be described by only two parameters, one giving the magnitude of the first constant and the other describing the ratio of successive constants. A number of systems approximate to this behaviour over a limited range of concentrations. Sillén himself was fastidious in never claiming that this type of model was a true description of the system and frequently emphasised that the interpretation was merely 'compatible with' the measurements available. The whole approach may seem unduly approximate; but to obtain even a general formula and two independent parameters from such intractable systems is a considerable achievement.

When a system of polynuclear complexes does not show any of the behaviour patterns described above, solution of equations (11.5) and (11.6) for values of β_{qp} virtually necessitates the use of a computer. But we must, of course, always remember that the reliability of stability constants (and even of the formulae of the complexes) is only as good as the measurements. The more complicated the system, the greater is the need both for a large number of measurements of M_t, L_t, [L] (and preferably also of [M]) and for a high standard of precision (see Fig. 11.5). With these criteria potentiometry is by far the most suitable technique, since it can be carried out using a wide range of values of M_t and introduces only one parameter, $E^\ominus$, which can be measured in the absence of complex formation.

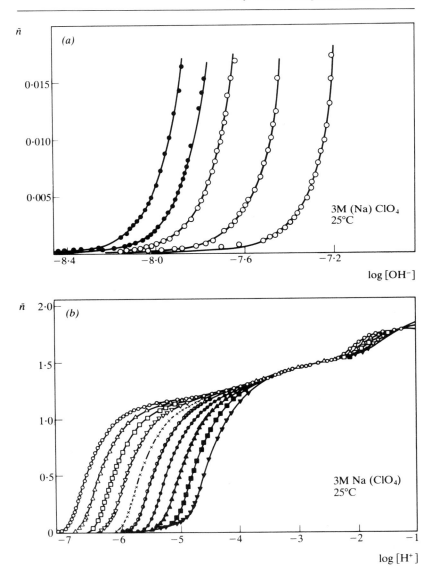

Fig. 11.5 Some high-precision work on difficult systems. (*a*) The hydrolysis of Ni^{2+}. The solutions are of low buffer capacity, and precipitation occurs at very low values of $\bar{n}$ ($\leqslant 0.02$). Measurements could only be made over a narrow range of pH, in which the main complex appears to be Ni$_4$(OH)$_4^{4+}$. (After K. A. Burkhov, L. S. Lilic and L. G. Sillén, *Acta Chem. Scand.*, **19**, 14 (1965).) (*b*) The protonation of the molybdate ion MoO$_4^{2-}$, which forms different series of complexes in different regions of pH. (After Y. Sasaki and L. G. Sillén, *Arkiv Kemi*, **29**, 253 (1968).)

It is sometimes possible to obtain the formula of a polynuclear complex by equilibrium centrifugation or by light scattering. Both methods give the least ambiguous results in monodisperse systems, i.e. those in which only one 'polymeric' species is present. However, even when several complexes of high molecular weight are formed, it may still be possible to establish a range within which the (appropriately weighted) average molecular weights lie.

11.2   Ternary complexes $M_q L_p X_x$

We may consider these as complexes of the species $M_q L_p$ with x additional groups, X, where X may be a second type of ligand L', the hydrogen or hydroxyl ion, or a second type of metal ion, M'. Except in these simplest of cases, systems which contain ternary, or 'mixed', complexes are very difficult to study, since the overall stability constants of a ternary complex is given by

$$\beta_{qpx} = \frac{[M_q L_p X_x]}{[M]^q [L]^p [X]^x} \tag{11.18}$$

and the elimination of the equilibrium concentration of the complex introduces the three variables [M], [L] and [X]. Most studies of mixed complexes have been confined to systems which are mononuclear in M; but interpretation of the measurements still presents problems, because the quantities α_0, $\bar{n}_L$ and $\bar{n}_X$ are all functions of the two variables [L] and [X]. If one or both ligands are non-basic, as in the many studies of metal halide and pseudo-halide complexes, the values of [L] and [X] may at least be varied independently. But if both types of ligand take part in protonic equilibria, the values of both [L] and [X] are affected by any change in pH. If possible, the stability constants of all binary complexes, such as $M_q L_p$, $M_q(OH)_p$ and $H_j L$, should be determined under conditions where no ternary species are formed; but in some studies of metal complexes of EDTA and its derivatives, the species ML and MHL are formed under such similar conditions that both stability constants must be obtained from the same set of measurements.

The general method for obtaining stability constants of ternary species is no different in principle from the familiar procedures for binary systems, outlined in Chapter 7. But, because of the complexity of the system, the measurements must be extremely precise and the range of total concentrations as wide as practicable; and since the calculations may be massive the use of a high-speed computer is highly desirable. It is not surprising that work on mixed complexes has so far been concentrated on systems which allow a number of simplifying assumptions.

Most ternary systems which have been studied have been mononuclear with respect to metal ion, and in some studies of the protonation of metal polyaminocarboxyls, measurements were made at such high metal ion concentrations that $M_t \gg L_t$ and that the presence of complexes containing more than one L group could be ignored. Alternatively, the total concentration of ligand may be in large excess of that of the metal ion, so that we may assume $[L] = L_t$. Complexes of a metal ion with two similar, non-basic ligands L and L′ (such as halide or pseudohalide ions) are often studied by varying the ratio $R = L_t/L_t'$ whilst keeping M_t and $(L_t + L_t')$ constants, e.g. by mixing equimolar solutions of ML_2 and ML_2' in different proportions. The equilibrium constant K_{disp} for the displacement reaction

$$ML_2 + ML_2' = MLL' \qquad (11.19)$$

can be obtained from the difference between some measured value of property, such as optical absorbency, or liquid–liquid distribution ratio, and the value predicted for mixing alone. The formation of mixed complexes often enhances the extraction of a metal into organic solvents. With mercury halides (see Fig. 11.6) the effect is slight, but the addition of organic bases to some metal complexes of TTA increases their extractability dramatically, by the so-called 'synergic effect' (see Fig. 11.7). This procedure for studying displacement reactions at constant $(L_t + L_t')$ is akin to Job's method of continuous variations (see sect. 10.1) in which $(M_t + L_t)$ is held constant. The stability constant of the mixed complex MLL′ is given by

$$\beta_{111} = K_{disp}(\beta_{101}\beta_{110})^{-\frac{1}{2}} \qquad (11.20)$$

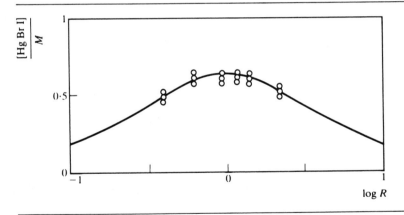

Fig. 11.6   The variation of [HgBrI]/Hg$_t$ with the ratio $R = Br_t/I_t$. The points refer to spectrophotometric measurements at different wavelengths, and the full line is calculated for $K_{disp} = 11\cdot9$. (After T. Spiro and D. N. Hume, *J. Amer. Chem. Soc.*, **83**, 4305 (1961).)

A similar, but more complicated, procedure has been described for studying the complexes MLL_2' and ML_2L' formed by mixing equimolar solutions of ML_3 and ML_3'.

Basic ligands present more of a problem. Fairly acidic solutions of copper(II) in the presence of two bidentate aminocarboxylate ions L and L′ may contain any of the species H^+, Cu^{2+}, L^-, HL, H_2L^+, CuL^+, CuL_2, L'^-, HL′, $H_2L'^+$, CuL'^+, CuL_2' and CuLL′. Even though stability constants of all

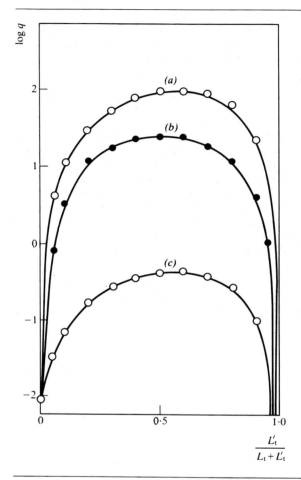

Fig. 11.7 The 'synergic' distribution of ^{60}Co between an aqueous solution of pH = 4·93 and cyclohexane containing TTA (L) and a base Q (L′) such that $(L_t + L'_t) = 0·02M$ and $L_t/L'_t = R$. The base Q is: (*a*) pyridine; (*b*) quinoline; (*c*) 2-methyl pyridine. (After H. M. N. H. Irving, D. Dyrssen, J. O. Liljenzin and J. Rydberg (eds), *Solvent Extraction Chemistry*, North-Holland, Amsterdam (1967).)

species except the last are known, extraction of the stability constants of the ternary complex CuLL′ from the data M_t, L_t, L'_t, H_t, and [H] is by no means easy.

Beautifully precise work on complexes of O-phosphoryl-serines and related ligands has been carried out by Österberg. His earlier measurements, on systems containing only mononuclear complexes, were analysed without a computer, despite the presence of a variety of species (e.g. in the Cu^{2+}, O-phosphoryl-ethano-amine system: ML, MLH, ML_2, and ML_2H). The value of [H] was measured over a wide range of ligand concentrations and the data [H], M_t, H_t, L_t subjected to a graphical treatment, involving interpolation and differentiation, and integration followed by more interpolation and differentiation, which would have reduced less worthy data to incoherence. Österberg has now computerised his treatment and extended it to include the formation of binary and ternary polynuclear species. But such sophistication is wasted unless the measurements are, like Österberg's, of superb quality.

Ternary, or even more complicated, species are formed in some aqueous solutions containing macromolecules and metal ions. Many proteins in aqueous solution bind a considerable number of hydrogen ions (see Ch. 6), some of which may well be replaced by other cations. Equilibria between proteins and metal ions have been studied much less than the analogous protonic equilibria, although the same models have been used for interpreting the measurements. The problems which bedevil the study of the protonation of macromolecules (Ch. 6) are equally, or even more, obtrusive in studies of the binding of metal cations; and it is harder still to study the binding of more than one type of cation.

The problem may be attacked most effectively by measuring [M] (e.g. by potentiometry or cation exchange) in addition to [H]. Measurements have usually been obtained either by equilibrium dialysis or by pH-titration. In the former, the metal ion is distributed between two solutions, one of which contains protein and is separated from the other by a membrane which is permeable to cations and small molecules, but not to proteins. After corrections for Donnan membrane effects, the free metal ion concentration in the protein solution is obtained from the total concentration of metal in the solution which contains no proteins. The concentration of metal bound to protein can then be obtained. The pH method involves determining the number of protons bound to a protein group both in the presence of a metal cation, and, as described in Chapter 6, in the absence of metal. The difference in the two proton-binding curves is attributed to binding of metal ion to protein.

There is clearly much scope for future work in this field: but the problems are considerable. All studies on ternary complex formation, and all on protein equilibria, need data of the highest precision, vigorously processed. For ternary protein complexes, the need for superb work is paramount, and the difficulty in doing it is almost overriding.

Equilibrium constants

Chapter 12

From measurements to parameters

We have so far taken it largely on trust that, given an adequate number of measurements and a suitable equation, values of equilibrium constants (and of other parameters such as extinction coefficients) 'can be obtained' provided that they can be held constant throughout the experiments. For systems of binary mononuclear complexes where quantities such as $\bar{n}$, α_c and $\mathscr{E}$ are functions of the single variable [L], the minimal requirement is that the number of pairs of measurements of $\bar{n}$, [L] and α_c, [L] or $\mathscr{E}$, [L] should not be less than the number of parameters we want to evaluate. So, in principle, we can obtain the N stability constants β_n for the system M–L from at least N pairs of values of $\bar{n}$ and [L] measured under conditions where activity coefficients, and hence stoichiometric equilibrium constants, do not vary. It is of course usual, and highly desirable, to make a much larger number of measurements than this minimal algebraic requirement. One would hardly wish to calculate the dissociation constant of a monobasic acid from a single pair of measurements; and the need for a generous number of data increases with the complexity of the system particularly if we are uncertain of the formula of all the complexes which are formed. As we saw in Chapter 7, values of the secondary variables $\bar{n}$ and α_c depend on two primary concentration variables if polynuclear binary complexes, or mononuclear ternary complexes, are formed. So a change from simple mononuclear systems to those of higher complexity causes an increase in the amount of information represented by each experimental point as well as in the total number of points needed.

In this chapter we shall merely outline some of the more acceptable ways in which to extract values of stability constants from experimental measurements. The choice of method naturally varies with the system; the best method for calculating the dissociation constant of a monobasic acid is unlikely to be the most suitable one for disentangling more complicated equilibria. But the criteria by which we may assess computational methods are common to all types of system.

To be acceptable, a method should give as clear an indication as possible which complexes are present. It should make full use of the experimental data (except for those points which are obviously 'dud') and be flexible enough to allow the weighting of measurements. A good method should show up systematic errors and allow corrections to be made for them. As well as yielding 'best' values of the parameters, the computational procedure should also give limits of error in these values. Finally, the method should be efficient in that, given a set of measurements, the maximum information should be extracted for the minimal necessary investment. This last quantity is a complex function of time, money, effort and boredom, with its exact composition depending on personal tastes and circumstances.

Two main approaches have been used. Data of high precision can in general be satisfactorily processed by electronic computation, although data which can be described by not more than two parameters can be treated equally well by graphical methods. The essence of either approach is the comparison of an experimental function, obtained from the measurements, with theoretical functions calculated by substituting values of the parameters into the appropriate equation. For example, an experimental formation curve $\bar{n}_{expt}(\log [L]_{expt})$ derived from measurements of M_t, L_t and $[L]$ may be compared with a number of theoretical formation curves $\bar{n}(\log [L])$ obtained by substituting different sets of values of β_n into equation (7.20). The most acceptable values of the stability constants are those used to calculate the function which best reproduces the experimental formation curve. The computer and the graph paper fulfil a number of common functions, in that they act as data stores; they can rapidly separate theoretical functions incorporating different values of (at least two) parameters; and they facilitate assessment of which theoretical function best represents the data. Both approaches involve subjective judgement both in deciding the relative importance of data obtained from different concentration ranges and in setting up criteria for the 'best fit' between the experimental and calculated functions.

Computers offer enormous advantages over graph paper, particularly for systems which involve more than two parameters (see sect 12.2). They free the researcher from the drudgery of manual arithmetic, although small doses of calculation are invaluable for giving him a 'feel' both for the function he is using and the peculiarities of the particular system. Thanks to the ubiquitous pocket calculator, acquisition of such insight can now be almost painless. Computerised methods are, of course, much quicker than graphical ones, provided that the program is already available and that no account is taken of the period spent understanding the program or waiting for computer time. If the results of one set of measurements on a two-parameter system are needed in order to plan the next experiment, a graphical method may prove more convenient than computerised processing; and it is cheaper by far.

Since many people absorb information more readily from a graph than from rows of figures, graphs are often more satisfactory than numerical print-out as a basis for discussion with those who lack the experimenter's familiarity with the problem. And graphs are almost essential for display purposes. When the measurements are processed by computer, it is often advantageous to prepare manual plots of important functions, such as the experimental, and 'best' theoretical, formation curves, unless a graphical print-out can be obtained.

It is sensible to process any exploratory set of measurements by manual calculation and graphical methods in order to get a general view of the behaviour of the system. We can detect the presence of polynuclear, or ternary, complexes in this way and we can learn much about mononuclear systems from the shape of the formation curve. Thus if plateaux are formed at integral values of $\bar{n}$, separated by x integers, we need only consider x equilibria in that concentration range. When the highest measured value of $\bar{n}$ lies on a plateau we therefore know the formula ML_N of the highest complex present. However, it is often impossible to obtain a complete formation curve, ending in a plateau. We may not be able to use sufficiently high values of [L] to convert all M into ML_N, or precipitation may occur when $\bar{n}$ is lower than N. Information about the number of complexes formed can, however, be obtained by comparing the shape of even a limited part of the formation curve with the theoretical curve for $N = 1$, and the family of steeper curves for systems in which $N = 2$ (see sect. 4.2). If the fragment of the formation curve is steeper than any of these normalised curves, the presence of higher complexes, such as ML_3, should be investigated.

Plots of formation curves are also useful for revealing errors. Random dud points obtrude and can be discarded. Plateaux at non-integral values of $\bar{n}$ (e.g. a formation curve starting at a constant value of $\bar{n} \neq 0$) indicate systematic errors which should be sought out and eliminated, probably by refining the analyses of the stock solutions. Systematic errors also cause formation curves to depart from their required symmetry which, for systems in which $N = 1$ or $N = 2$, is rotational about the mid-point value of $\bar{n}$.

If preliminary graphical treatment indicates that the measurements are free from obvious errors and that only one, or two, binary mononuclear complexes are formed, we may obtain definitive values of equilibrium constants either by graphical or computation methods. More complicated systems are probably better handled by both methods, the final values being obtained computationally.

12.1 Graphical methods

Graphs can, of course, represent both straight lines and curves. There is a bias in our early training towards drawing straight lines through sets of points, perhaps on account of algebraic simplicity or of the frequency of straight edges in our culture (and the consequent availability of rulers).

In one-parameter systems, data may be represented by straight lines in which either the slope, or the intercept, is fixed. For systems in which $N = 1$, we may define a quantity

$$Q_1 = \frac{\bar{n}}{1 - \bar{n}} = \frac{\alpha_1}{1 - \alpha_1} = \frac{\alpha_1}{\alpha_0} = \frac{1 - \alpha_0}{\alpha_0} = \beta_1[L] \tag{12.1}$$

We may therefore obtain the value of β_1 as the slope of the line drawn through the origin and the points Q_1, [L]. Alternatively, the value of β_1 may be obtained as the intercept of the horizontal line $Q_1[L]^{-1}$, [L]; or the value of $\log \beta_1$ as the intercept of the line drawn with unit slope through the points $\log Q_1$, $\log [L]$. Since the value of Q_1 $(= \bar{n}/(1 - \bar{n}))$ may be subject to gross error when $\bar{n} \sim 0$, or when $\bar{n} \sim 1$, only points within the range $0\cdot05 \leqslant \bar{n} \leqslant 0\cdot95$ should be used.

For systems in which $N = 2$, we may write equations (7.20) and (7.24) in linear form to give

$$Q_2 = \frac{\bar{n}}{(1-\bar{n})[\mathrm{L}]} = \beta_1 + \frac{(2-\bar{n})[\mathrm{L}]\beta_2}{(1-\bar{n})} \tag{12.2}$$

$$Q_2' = \frac{1-\alpha_0}{\alpha_0[\mathrm{L}]} = \beta_1 + \beta_2[\mathrm{L}] \tag{12.3}$$

Values of β_1 and β_2 are then obtained as the intercept and gradient of a plot of Q_2 against $(2-\bar{n})[\mathrm{L}]/(1-\bar{n})$ or of Q_2' against $[\mathrm{L}]$. Similar expressions can represent data α_1, $[\mathrm{L}]$ and α_2, $[\mathrm{L}]$ in linear form. But in all cases, points in the region of integral values of $\bar{n}$ or α_c lead to ill-conditioned equations and must be discarded. Linear methods suffer from two further disadvantages:

(i) since the functions $Q_1, Q_2, Q_3, \ldots$ are rather widely removed from the secondary variables $\bar{n}$ and α_c, linear plots do not give a very clear indication of systematic errors; and
(ii) since $[\mathrm{L}]$ is usually varied over a number of orders of magnitude, the points tend to be bunched together.

For these reasons, it is often preferable to obtain values of the parameters by non-linear curve-fitting (see below).

We cannot, of course, obtain values of more than two constants from a single linear plot. However, systems containing three or more complexes have been treated by successive linear extrapolation. Suppose that values of α_0 and $[\mathrm{L}]$ have been measured for a system containing N complexes. At low free ligand concentrations, only the first one or two complexes will be formed and so values of β_1 and β_2 may be obtained as the limiting intercept and gradient, as $[\mathrm{L}] \to 0$, of the plot of $(1-\alpha_0)/\alpha_0[\mathrm{L}]$ against $[\mathrm{L}]$ (cf. equation (12.1)). We may use the value of β_1 to calculate the function

$$Q_3' = \frac{Q_2' - \beta_1}{[\mathrm{L}]} = \beta_2 + \beta_3[\mathrm{L}] + \beta_4[\mathrm{L}]^2 + \cdots \tag{12.4}$$

As $[\mathrm{L}]$ tends to zero, the function $Q_3'([\mathrm{L}])$ tends to a straight line. The intercept provides a check on the value of β_2 while the gradient gives the value of β_3. This procedure may, in principle, be repeated until all N stability constants have been obtained. Although errors accumulate with each successive extrapolation, this trouble can sometimes be alleviated by carrying out the analogous extrapolation to high free ligand concentrations, using plots of functions such as

$$\frac{(\alpha_0-1)}{\alpha_0[\mathrm{L}]^N} = \beta_N + \beta_{N-1}[\mathrm{L}]^{-1} + \cdots \tag{12.5}$$

against $[\mathrm{L}]^{-1}$. Extrapolations to $[\mathrm{L}]^{-1} \to 0$ give values of stability constants β_n which decrease in reliability as n decreases. Values of β_n for low values of n

may therefore be obtained by extrapolation to $[L] \to 0$ and those for higher complexes by extrapolation to $[L]^{-1} \to 0$. However, even with this refinement, the successive extrapolation method is unsatisfactory and, if used at all, the linear plots are best restricted to systems in which not more than two complexes are formed.

Values of parameters for systems in which $N = 1$ or $N = 2$ may often be obtained very satisfactorily by non-linear curve-fitting. We have seen that, if a single complex ML is formed, the experimental plots of $\bar{n}$ ($=\alpha_1$) or α_0 against log $[L]$ are of unique shape, identical to that of the theoretical plots of

$$\bar{n} = \frac{l}{1+l} \tag{12.6}$$

and

$$\alpha_0 = \frac{1}{1+l} \tag{12.7}$$

against log l, calculated using the normalised variable

$$l = \beta_1[L] \tag{12.8}$$

The value of β_1 may readily be obtained by curve-fitting. The experimental points are first plotted as $\bar{n}$ or α_0 against log $[L]$. The theoretical curve of $\bar{n}$ or α_0 as a function of log l is then calculated from equation (12.6) or (12.7) and plotted on the same scale. One curve is placed over the other so that their vertical axes (representing $\bar{n}$ or α_0), are coincident. One graph is then moved horizontally over the other until the best possible fit is obtained between the experimental and theoretical curves. In this position the origin of the normalised curve coincides with the point $(0, -\log \beta_1)$ of the experimental plot. A draughtsman's light-table is ideal for curve-fitting, but a large window, or a sheet of transparent graph paper, is an acceptable substitute. For a one-parameter system of the type we have discussed, curve-fitting is an excellent method which enables the experimenter to fit all data simultaneously whilst allowing him to judge points in some regions to carry greater weight than others. Random error shows up as scatter while systematic error causes the experimental curve to depart from the 'theoretical' shape. The effects of some common errors are shown in Fig. 12.1. The normalised curve can be calculated very rapidly and, once plotted on a suitable scale, is scarcely less convenient a template than a ruler.

A similar curve-fitting procedure can be used to obtain values of two parameters for a system containing the single complex ML studied by measuring a property (such as distribution ratio, or spectrophotometry of one absorbing species) which is proportional to α_0 or α_1. Suppose, for example, we study liquid–liquid partition in a system where the sole complex, ML, is uncharged. Since

$$q = P_1\alpha_1 = \frac{P_1\beta_1[L]}{1+\beta_1[L]} \tag{12.9}$$

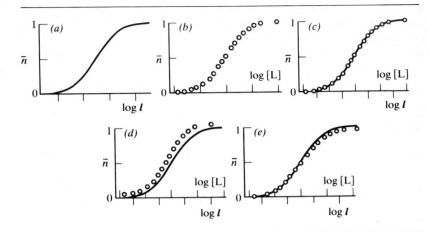

Fig. 12.1 Formation curves for single complex ML: (*a*) the theoretical plot of $\bar{n}$ against log l (equation (12.6)); (*b*) imaginary data $\bar{n}$(log [L]); (*c*) best fit of (*a*) and (*b*); (*d*) effect of systematic analytical error in L_t; (*e*) effect of systematic analytical error in the ratio L_t/M_t. (From F. J. C. Rossotti and H. Rossotti, *The Determination of Stability Constants*, McGraw-Hill, New York (1961).)

both variables may be normalised. The curve log q, (log [L]) is therefore of unique shape, identical to that of the normalised curve log q(log l) where

$$\log q = \log qP_1^{-1} = \log \frac{l}{1+l} \tag{12.10}$$

and the normalised variable l is again defined by equation (12.8). The experimental points log q, log [L] are superimposed on the normalised curve in the position of best fit compatible with the axes of the two plots being parallel. The two parameters can then be obtained from the coordinates, log P_1 and $-\log \beta_1$, of the point on the experimental plot which coincides with the origin of the normalised curves (see Fig. 12.2).

When two complexes are formed, the shape of the plot of $\bar{n}$ or α_c against log [L] depends on the relative stability constants (see sect. 4.2). But curve-fitting methods may nonetheless be used. It is advisable to calculate a family of normalised curves $\bar{n}$(log l)ρ using equation (4.33) and a few, widely spaced values of $\rho = (K_1/K_2)^{\frac{1}{2}}$, e.g. 10^2, 10, 1 and 0. These plots will, of course, be analogous to the functions $\bar{j}$(log h) shown in Fig. 4.2. The experimental data $\bar{n}$, log [L] are first superimposed with coincident ordinates on the theoretical curves, just to check that the experimental curve fits in with the set. If it does, there are a number of graphical methods for determining the values of β_1 and β_2. One of the most convenient is the projection-strip method. The procedure is exactly analogous to that described in sect. 4.5 for obtaining the formation constants of a dibasic acid from measurements of $\bar{j}$ and log [H]. The experimental formation curve is projected on to the log [L] axis to form a strip consisting of a number of values of log [L] corresponding to certain fixed

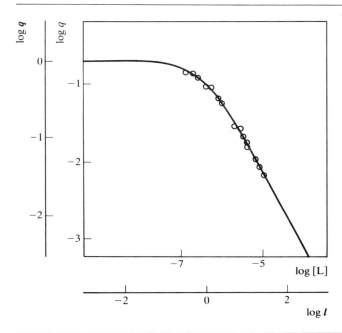

Fig. 12.2 Theoretical curve log q (log l), equation (12.10), superimposed on hypothetical data log q, log [L] in the position of best fit, corresponding to $P_c = 20$ and log $\beta_1 = 6\cdot4$. Note that a value of P_c can be obtained using data from values of [L] for which $q < P_c$.

values of $\bar{n}$, e.g. $\bar{n} = 0\cdot1$, $0\cdot2$, $0\cdot3$. Realistic limits of error in log [L] can also be represented on the strip. Using the same values of fixed $\bar{n}$, and the same scale for the abscissa, a set of normalised curves log $\rho(\log l)_{\bar{n}}$ is calculated with the help of equation (4.33). The strip is then superimposed on the normalised curves, parallel to the log l axis and in the position of best fit (cf. Fig. 4.3). The value of log ρ is obtained from the ordinate, while that of $\frac{1}{2}$ log β_2 is the value of the abscissa of the experiment strip at its intersection with the vertical line log ρ, 0 representing the normalised function for $\bar{n} = 1$. Once the set of normalised curves has been prepared, the projection-strip method is quick, easy and reliable. However, unlike the simple curve-fitting treatment of systems in which $N = 1$, it involves the processing of smoothed functions rather than of the raw data.

Methods similar to those described for processing data $\bar{n}$, [L] may also be used for treating measurements of α_c, [L]. Properties which are proportional to α_c may yield three parameters if measurements have been made over a sufficiently wide range of concentrations. For example, the distribution ratio q for a system in which ML partitions may be expressed as

$$q = P_1\alpha_1 \qquad\qquad (12.11)$$

whence

$$\log \boldsymbol{q} = \log q - \log P_1 = \frac{\rho l}{1 + \rho l + l^2} \qquad (12.12)$$

If [L] can be varied sufficiently widely, it may be possible to obtain values of P_1 and β_2 by fitting the experimental plot of $\log q$ against $\log [L]$ to the asymptotes of the set of normalised curves $\log \alpha_1 (\log [L])$. Values of α_1, [L] may then be calculated and treated to give the value of ρ (and a value of β_2 which should, of course, agree with that obtained previously).

When three or more complexes are present, curve-fitting cannot be applied simultaneously to data obtained over the whole concentration range, but must be restricted to regions where there are not more than two (at least predominant) complexes. Preliminary values of stability constants can conveniently be obtained in this way, but refined values of constants for systems which contain more than two complexes are normally better treated by computer than by graphical methods.

12.2   Computerised methods

Computational techniques, like graphical ones, vary with the complexity of the system. They are most valuable for highly complex systems provided, of course, that the experimental measurements are of sufficiently high quality. Since there have been several useful reviews of the applications of computers to equilibrium (and to other types of) chemistry, we shall treat the subject only very briefly.

As with graphical techniques, computerised methods find the 'best' set of stability constants by comparing the experimental data with analogous quantities calculated by assigning sets of values to the constants (and sometimes also to other parameters such as partition coefficients and extinction coefficients). The final values are those which best fit the data. The data are processed so that each point, i, gives a quantity y_i which may be expressed in terms both of the parameters β_x and of other quantities $(x_{i1}, x_{i2}, \ldots, x_{ij})$ which can be calculated from the experimental data. For example, a system of binary mononuclear complexes may be treated using equation (7.20) in the form

$$\bar{n} = \sum_{n=1}^{N} (n - \bar{n})[L]^n \beta_n \qquad (12.13)$$

When values of M_t, L_t and [L] have been measured, they may be used to calculate values of

$$\bar{n}_i = y_i \qquad (12.14)$$

and of the terms x_{in}, viz:

$$(1 - \bar{n})[L] = x_{i1}; \qquad (2 - \bar{n})[L]^2 = x_{i2}; \qquad \text{etc.} \qquad (12.15)$$

For the point i, equation (12.13) then becomes

$$y_i = x_{i1}\beta_1 + x_{i2}\beta_2 + \cdots \qquad (12.16)$$

Exactly analogous equations, containing different values of $\bar{n}$ and of the various terms $(n - \bar{n})[L]^n$, may be written for each of the other experimental points. A set of calculated values of y_{ic}, y_i, may be obtained by combining the values of x_{ij} used in equation (12.15) with a set of values of β_n. Good agreement between experimental and calculated values of y at the point i leads to only small differences between y_i and y_{ic}. Since the residual

$$U_i = y_i - y_{ic} \qquad (12.17)$$

may be positive or negative, we use a low value of U_i^2 (rather than of U) as criterion of good fit, and attempt to choose values of parameters which yield small squared residuals over the whole range of i. But because changes in the parameters may affect the value of y_{ic} more for points in some concentration ranges than in others, many researchers introduce different weighting factors, w_i, for each point. So the criterion of best fit is that the sum

$$U = \sum_i w_i U_i^2 = \sum_i w_i (y_i - y_{ic})^2 \qquad (12.18)$$

of the weighted squares of the residuals for all points, should be minimal.

In principle, there is little difficulty in the computerised solution of 'linear' equations such as (12.16) in which all parameters appear in separate terms and are raised only to the first power. No preliminary estimate of the values of the parameter is involved. Difficulties arise in practice, however, because both y_i and x_i are subject to experimental error which is, moreover, correlated. For example, an error in $[L]_i$ will cause errors in both $\bar{n}$ and in all the terms $(n - \bar{n})[L]_i^n$. The effect of correlated errors may influence the choice of weighting factors w_i.

When polynuclear, or mixed, complexes are formed, the relationship between y_i and x_i may involve parameters raised to powers greater than unity or products of two or more parameters. Solution of equations analogous to (12.16) then requires preliminary estimates B_x of the parameters β_x. These initial values may be obtained by graphical methods and are refined by a computerised technique which considers the effect on y_{ic} of small changes in the parameters B_x. The partial derivatives

$$d_{ix} = \frac{\partial y_{ic}}{\partial \log B_x} \qquad (12.19)$$

can be used in a set of linear equations which can be solved to give the parameters, $\Delta \log B_x$, representing changes which must be made, in turn, to each value of $\log B_x$ in order to minimise U. When successive increments become vanishingly small for each value of x, the summation U will have the lowest possible value and the final values of B_x represent an acceptable set of the parameters β_n. The speed with which convergence is achieved depends on the accuracy of the initial estimates, B_x. If these values differ widely from the correct ones, they may never converge. Convergence may also be very slow if a change in a value of one parameter necessitates appreciable adjustments in the values of others. Not surprisingly, this so-called covariance is aggravated by the presence of several polynuclear species, or by inadequacies in the quality, or quantity, of the data.

For studies of polynuclear hydrolysis of metal ions, Sillén and his co-workers used programs in which the position of best fit was found by means of an expression which contained second derivatives of the functions $y_{ic}(\log B_x)$ as well as the first derivatives d_{ix}. This procedure facilitates convergence even with poor preliminary estimates of the parameters. The most sophisticated of these programs, LETAGROP VRID, can handle data which contain parameters, such as extinction coefficients, in addition to the stability constants. Less traditionally, systematic errors, e.g. in the composition of stock solutions, are also treated as adjustable parameters. Problems arising from covariance can be reduced by making related changes in $\log B_x$ rather than varying the values independently.

Although LETAGROP VRID is the most comprehensive approach to the computing of stability constants, it is expensive on time; both on computer running-time and on the time required for the researcher to learn to use, and to adapt, the program. So sophisticated a procedure is probably best restricted to large problems and then only if the data are of very high quality.

Since all computerised treatments are fundamentally methods for testing how well various theoretical models fit the data, we need first to select those models which we wish to test. It is normally possible to use graphical methods to establish the formulae and the approximate stability constants of the one or two predominant species, together perhaps with approximate values of the empirical formula (e.g. of the ratio q/p for the species M_qL_p) of the minority ones. Models containing the established predominant species and various plausible minority ones are tested in them. Baes and Mesmer recommend first using models with the fewest possible number of species and considering more only if no acceptable fit can be found. If more than one possible model emerges, the preferred minor species is that which gives the lowest value of U (see Fig. 12.3). Sillén's school used the converse procedure of considering that a system may contain a number of plausible species and then rejecting those which are found to be of very low stability. Thus a species may be deemed to make a negligible contribution to the system if the value of the stability constant does not exceed its uncertainty by a specified factor (e.g. 1·5).

Not surprisingly, Sillén's approach gives rise to a greater variety of minority species than does Baes and Mesmer's, and the two methods lead to slight differences in the stability constants (but, reassuringly, to no difference in the formulae) of the predominant complexes. These minor uncertainties serve to emphasise, again, that the 'best' interpretation is not necessarily the 'right' one. Stability constants obtained by any method of calculation, however automated, can never rise above the status of values 'compatible with' the experimental data. And, the more extensive and precise the data, the greater is the chance that the interpretation is 'right'.

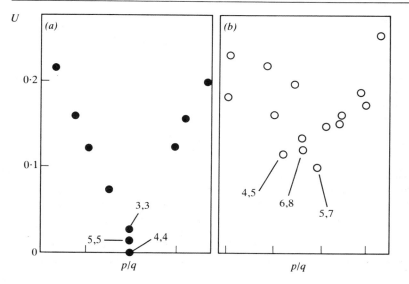

Fig. 12.3 Use of the residual U, equation (12.17), to determine the formula of a complex M_qL_p. (a) hypothetical data for a system containing the unique complex M_4L_L; (b) data for the hydrolysis of Be^{2+}, resulting in the formation of Be_2OH^{3+}, $Be_3(OH)_3^{3+}$ and a minor species $Be_q(OH)_p^{(2q-p)+}$. Only the three most likely sets p, q are labelled in either diagram. (After C. F. Baes and R. E. Mesmer, *The Hydrolysis of Cations*, Wiley, New York (1976).)

Once values of the stability constants have been obtained, by whatever method, they should always be used to back-calculate the original measurements, so that, for example, values of $\bar{n}_i$, [L] can be compared with $\bar{n}_{ic}$, [L]. For many methods of calculation, such as computerised procedures and non-linear curve-fitting for systems in which $N < 2$, such comparison is, of course, inherent in the method. Visual comparison is often instructive and results may conveniently be displayed as a graph showing both the experimental points $\bar{n}(\log[L])$ or $\alpha_c(\log[L])$ and the theoretical curves calculated using the 'best' set of parameters.

Limits of error in the parameters should always be obtained, and quoted, whether they be standard deviations calculated by a rigorous computerised statistical treatment or visual estimates obtained in graphical methods.

Chapter 13

Making use of stability constants

A glance at the compilation (p. 189, ref. 43) of stability constants may stimulate speculation on the human resources which have so far been spent on measuring equilibrium constants in ionic solutions. Values of acid dissociation constants were first reported in 1889. Although most stability constants of metal complexes have appeared since the publication in 1941 of J. Bjerrum's thesis on metal ammines, metal complex equilibria have been studied intermittently throughout the century, one of the earliest workers, Bodländer, being killed in a car-crash in 1904. The hours and ergs expended per value measured vary enormously with the complexity of the system and the dedication and efficiency of the research team; but whatever the system, they can be guaranteed to be far in excess of the orignal estimate. And since equilibrium chemistry of often a slow and painstaking business, it is proper to ask whether the results are worth the effort. There are, of course, personal reasons for measuring equilibrium constants, akin to those which lead us to undertake any other activity whether it be a routine job, collecting stamps, making lace or climbing Everest. But of what use, if any, are the equilibrium constants so obtained?

There are a number of positive answers to this question, all coming under the general head of increasing our understanding of chemical behaviour. We have seen that, for binary mononuclear complexes, a value of the stability constant significantly greater than $0 + x$ (where x is the expected error) may be taken as strong evidence that the species can be formed in appreciable concentrations. In more complicated systems, however, such definite interpretation may not be possible. When several polynuclear complexes coexist, the measurements may sometimes be represented almost equally well by two or more sets of parameters, each set therefore being merely 'compatible with' the presence of certain species rather than providing unequivocal evidence for their existence.

Most determinations of equilibrium constants aim not merely to establish the formula of a complex, but to obtain a numerical value of a constant which is required, often for one of the following purposes:

1. the calculation of the concentrations of various species in a solution;
2. the calculation of such equilibrium constants as β_i^H, which are often needed for calculating stability constants of metal complexes;

3. the calculation of other thermodynamic functions, in particular enthalpies and entropies of complex formation;
4. the calculation of intensive factors, e.g. the processing of absorptiometric measurements to yield extinction coefficients;
5. discussion of the various factors which govern strengths of acids and stabilities of complexes.

These various uses of stability constants will be discussed more fully below.

13.1 Calculation of equilibrium concentrations

Once the stability constants have been determined, we can, in principle, calculate the equilibrium concentration of any species which is present in solution. We shall see in sect. 13.7 that a knowledge of equilibrium concentrations is essential for an interpretation of many aspects of the behaviour of a solution, e.g. its spectroscopic and thermochemical properties. Such knowledge also facilitates the design of new separative and analytical procedures and may enable us to predict: the pH at which there is the maximal separation of two metal complexes by solvent extraction; the type of masking agent which will best hold one ion in a solution from which we can quantitatively precipitate another; the conditions under which a routine colorimetric analysis should be carried out; and so on. Examples are provided by the numerous large books on the theory of classical analysis.

The help provided by ionic equilibrium constants is in no way restricted to the more traditional aspects of chemistry. The impact of equilibrium chemistry on life sciences ranges from enabling a biochemist to prepare a buffer of suitable pH and ionic strength to stimulating him to new tactics in attempting to understand nerve action, cell membranes, the importance of trace metals and the growth of sea-shells. Oceanographers, too, have latterly become interested in the detailed composition of the vast solution with which they work. A knowledge of equilibrium chemistry might increase detailed understanding and further the development of technological procedures in such crafts as photographic processing and museum conservation work. But even if we can put forward no practical reason for calculating the concentrations of the species present at equilibrium, there is a strong, albeit non-utilitarian, case for 'knowing what's going on'.

Binary mononuclear systems are, not surprisingly, the easiest to handle. Valuable insight into a particular system may be obtained by calculating the distribution of central groups between the various complexes as a function of the free ligand concentration. If the value of the free ligand concentration in a particular solution is not known, it can be obtained by solving the appropriate equations, e.g. (7.11) and (7.14), which express the overall concentrations in terms of the stability constants and free ligand concentration (see sect. 13.3).

Variation of α_c, $\bar{n}$ and $R_{c,r}$ with [L]

For simplicity, the discussion will be mainly limited to metal complexes, but the methods described can, of course, equally well be applied to the successive protonation of a base.

The distribution of a metal ion among its complexes may readily be calculated from equation (7.24). Since the fraction of the total metal ion in the form of the particular species ML_c (where $0 \leqslant c \leqslant N$) is given by

$$\alpha_c = \frac{[ML_c]}{\sum [ML_n]} = \frac{\beta_c [L]^c}{\sum \beta_n [L]^n} \tag{13.1}$$

the total concentration of ML_c is

$$[ML_c] = \alpha_c M_t \tag{13.2}$$

If we need a value, $\bar{n}$, for the average number of ligands bound to each metal ion, we may combine equations (7.20) and (7.24) to give

$$\bar{n} = \frac{\sum n\beta_n [L]^n}{\sum \beta_n [L]^n} = \alpha_1 + 2\alpha_2 + 3\alpha_3 + \sum c\alpha_c \tag{13.3}$$

It is sometimes useful to calculate merely the ratio of the concentration of each complex ML_c to that of a particular reference complex ML_r. The value $R_{c,r}$ of this ratio may be calculated very easily as

$$R_{c,r} = \frac{[ML_c]}{[ML_r]} = \frac{\alpha_c}{\alpha_r} = \beta_c \beta_r^{-1} [L]^{c-r} \tag{13.4}$$

13.2   Display of results

Values of α_c, [L] may conveniently be displayed as plots of either α_c, or of $\sum_0^c \alpha_c$, against $\log [L]$ (see, e.g., Figs 13.2–13.4 and 13.6). The set of curves $\alpha_c (\log [L])_{c=0...N}$, which will be termed a *distribution plot*, demonstrates how the concentration of any of the intermediate complexes passes through a maximum as [L] is increased. Such plots show clearly the range of free ligand concentration over which each complex should be taken into account. The set of curves $\sum_0^c \alpha_c (\log [L])_{c=0...N}$ does not show up maxima in the values of α_c; but since the left-hand curve in the family represents α_0, the second $(\alpha_0 + \alpha_1)$, and so on, a vertical line drawn at a particular value of [L] gives the value of α_0 where it cuts the lowest curve, and that of α_1, as the vertical distance between the first and second curves. Values of α_c for higher complexes may be obtained similarly. Since these sets of curves are useful for obtaining a direct measure of the percentage of each complex present at a particular value of [L], we shall call them *percentage plots*.

Data α_c, [L] are sometimes displayed in the form of a *log–log plot* in which $\log \alpha_c$ is plotted against $\log [L]$. Although such plots are seldom applied to metal complex formation they are very useful for obtaining the value of pH in complicated buffer mixtures (see sect. 13.4).

Values of $\bar{n}$ are usually plotted against $\log[L]$ to give a *formation curve*. Isolated values of $\bar{n}$, $[L]$ give little insight into the state of a solution; a value of $\bar{n} = 1$ can mean that all the metal ion is in the form of ML ($\alpha_0 = 0$, $\alpha_1 = 1$, $\alpha_2 = 2, \ldots$) or that half of it is in the form of ML_2 while the rest is uncomplexed ($\alpha_0 = 0.5$, $\alpha_1 = 0$, $\alpha_2 = 0.5$, $\alpha_3 = 0, \ldots$). But the shape of the formation curve can nonetheless provide strong clues about the distribution of the complexes (see below and sect. 4.2 and 5.1). Since formation curves provide less visual information than distribution plots it might seem pointless to calculate them. However, the great majority of determinations of stability constants involve measuring values of $\bar{n}$ and $[L]$. And, once values of β_n have been obtained, it is essential to check that there is satisfactory agreement between the experimental values of $\bar{n}$ and $[L]$ and the theoretical values of $\bar{n}$ calculated for the same values of $[L]$ by substituting the stability constants into equation (13.3). Calculated formation curves may also be used to determine the value of free ligand concentration, and hence the distribution of complexes, in solutions in which only the total concentrations are known (see sect. 13.3).

The easiest way of representing equilibria is to plot $\log R_{c,r}$ against $\log[L]$ to give an *equilibrium ratio diagram*. Since, from equation (13.4),

$$\log R_{c,r} = (\log \beta_c - \log \beta_r) + (c - r) \log[L] \tag{13.5}$$

all functions are straight lines of integral slope, and the diagram may readily be constructed. At any value of $[L]$, the predominant species is that with the highest value of $R_{c,r}$. The ratio of the concentration of ML_c to that of any other complex ML_n is given by $[ML_c]/[ML_n] = R_{c,r}/R_{n,r}$. The function $\log R_{r,r}(\log[L])$ for the reference species is, of course, a horizontal line passing through the origin.

We shall now examine the form of these plots in rather more detail, dealing first with binary systems in which one, two or 'several' mononuclear complexes are formed, and then with binary polynuclear complexes, and ternary systems.

A single complex ML

All the diagrams we have mentioned are of unique shape and vary (with β_1) only in their position on the $\log[L]$ axis. We may demonstrate this by replacing the product $\beta_1[L]$ by the normalised variable l. Then for $N = 1$, equations (13.1) and (13.3) give

$$\alpha_0 = \frac{1}{1+l} \tag{13.6}$$

$$\alpha_1 = \frac{l}{1+l} = \bar{n} \tag{13.7}$$

and, for the equilibrium ratios (13.4), we have

$$R_{0,1} = \frac{1}{l}; \qquad R_{1,1} = 1; \qquad R_{0,0} = 1; \qquad R_{1,0} = l \tag{13.8}$$

Thus the ordinates of all the plots we have mentioned depend only on the normalised variable I and so are of unvarying shape. Moreover, since

$$\log I - \log [\text{L}] = \log \beta_1 \qquad (13.9)$$

that value of β_1 governs the position of the experimental function on the $\log [\text{L}]$ axis.

The various types of plot are shown in Fig. 13.1. It is seen that the distribution plots (Fig. 13.1(a)) of α_0 and α_1 against $\log [\text{L}]$ intersect at the point where $\alpha_0 = \alpha_1 = 0.5$ and $\log I = 0$, or $\log [\text{L}] = -\log \beta_1$. This is of course analogous to the familiar relationship (3.28) which states that, for monobasic acids, $[\text{HA}] = [\text{A}]$ when $\text{pH} = \text{p}K$. Of the two percentage plots (Fig. 13.1(b)), the left-hand one is of course identical to the distribution plot $\alpha_0(\log [\text{L}])$ for uncomplexed metal ion; and since $(\alpha_0 + \alpha_1)$ is always unity when $N = 1$, the second is the horizontal line representing $\sum_0^c \alpha_c = 1$. The ratio of the distance below the sigmoid curve (to $\sum_0^c \alpha_c = 0$) to that above it (to $\sum_0^c \alpha_c = 1$) gives the ratio $[\text{M}]/[\text{ML}]$ which again becomes unity when $\log [\text{L}] = -\log \beta_1$. When $N = 1$, the formation curve $\bar{n}(\log [\text{L}])$ is identical to the distribution curve $\alpha_1(\log [\text{L}])$ for the first complex so that $\bar{n} = 0.5$ when $\log [\text{L}] = -\log \beta_1$.

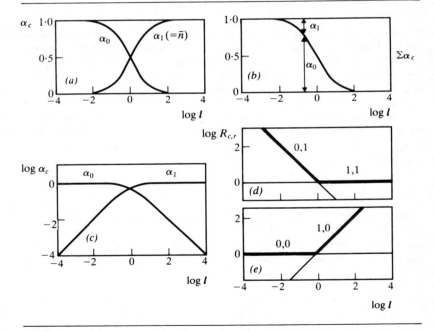

Fig. 13.1 The formation of a single complex ML. (a) Distribution plots: α_0 and α_1 ($=\bar{n}$) as functions of $\log I$ ($=\log \beta[\text{L}]$). (b) Percentage plot, $\alpha_0(\log I)$. (c) Log–log plots. Log α_0 and log α_1 as function of log I. (d) Equilibrium ratio diagram. Log $[\text{ML}_c]/[\text{ML}]$ as functions of log I. (e) Equilibrium ratio diagram. Log $[\text{ML}_c]/[\text{ML}]$ as functions of log I.

Plots with logarithmic ordinates are very easy to construct, particularly when a single complex is formed. At low values of [L] ($l \ll 1$), log–log plots (Fig. 13.1(c)) take the form of the two straight lines:

$$\log \alpha_0 = 0 \tag{13.10}$$

and

$$\log \alpha_1 = \log \beta_1 + \log [L] \tag{13.11}$$

while at high values of [L] ($l \gg 1$) we have the two straight lines:

$$\log \alpha_0 = -\log \beta_1 - \log [L] \tag{13.12}$$

and

$$\log \alpha_0 = 0 \tag{13.13}$$

The lines are curved in the region where both species exist at appreciable concentration and intersect at the point where $\log \alpha_0 = \log \alpha_1 = \log 0 \cdot 5$ and $\log [L] = -\log \beta_1$. The equilibrium ratio plots are even easier to draw since they are made up entirely of straight lines (regardless of the value of N). If ML is taken as the reference complex (Fig. 13.1(d)), the plot of $\log R_{0,1}$ against $\log [L]$ is also a straight line through the point $(0, -\log \beta_1)$ but with a slope of -1, cf. equation (13.4). Alternatively, if M is chosen as reference species (Fig. 13.1(e)) the line $\log R_{1,0}(\log [L])$ is a straight line of unit slope intersecting the horizontal line $\log R_{0,0} = 0$ at the point where $\log [L] = -\log \beta_1$.

Two complexes ML and ML$_2$

Diagrams representing the formation of two complexes show more variety than those for systems in which $N = 1$ because many of the functions are not of unique shape. The form of the various plots may again be conveniently discussed in terms of a normalised free ligand concentration. For systems in which $N = 2$, we shall use the normalised variable

$$\mathbf{L} = \beta_2^{\frac{1}{2}}[L] \tag{13.14}$$

The term $\beta_1[L]$ then becomes $\beta_1\beta_2^{-\frac{1}{2}}\mathbf{L}$, or $\rho\mathbf{L}$, where

$$\rho = \beta_1\beta_2^{-\frac{1}{2}} = (K_1/K_2)^{\frac{1}{2}} \tag{13.15}$$

We may now write the familiar expressions for α_c and $\bar{n}$ in terms of $\mathbf{L}$ and ρ. From equations (13.1), (13.14) and (13.15) we obtain

$$\alpha_0 = \frac{1}{1 + \rho\mathbf{L} + \mathbf{L}^2} \tag{13.16}$$

$$\alpha_1 = \frac{\rho\mathbf{L}}{1 + \rho\mathbf{L} + \mathbf{L}^2} \tag{13.17}$$

$$\alpha_2 = \frac{\mathbf{L}^2}{1 + \rho\mathbf{L} + \mathbf{L}^2} \tag{13.18}$$

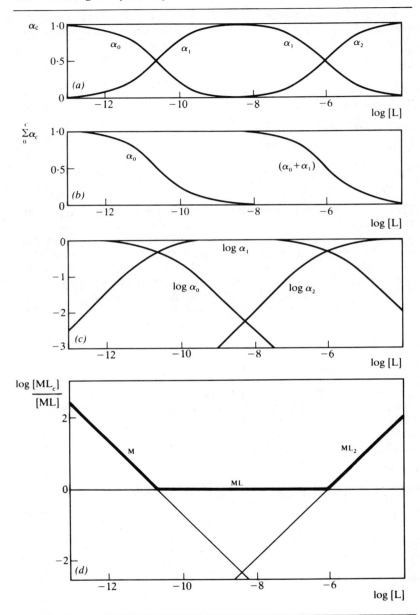

Fig. 13.2 The system Cu^{2+}–iminodiacetate $HN(CH_2COO^-)_2$ in which $N = 2$, $\log K_1 = 10 \cdot 63$ and $\log K_2 = 6 \cdot 05$. (*a*) Distribution plots $\alpha_c(\log[L])$. (*b*) Percentage plots $\sum_0^c \alpha_c(\log[L])$. (*c*) Log–log plots $\log \alpha_c(\log[L])$. (*d*) Equilibrium ratio diagram, $\log[ML_c]/[ML](\log[L])$.

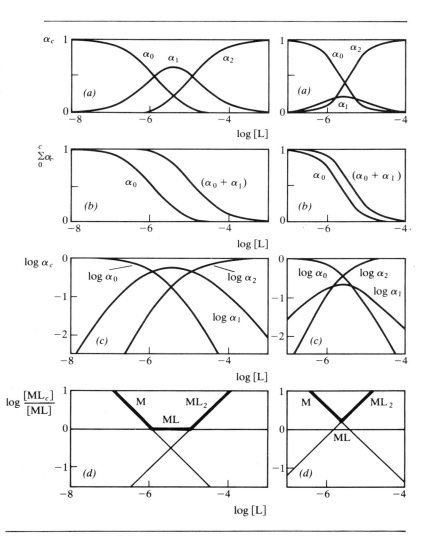

Fig. 13.3 The system Cu(I)–NH₃ in which $N = 2$, $\log K_1 = 5\cdot93$ and $\log K_2 = 4\cdot93$. (*a*) Distribution plots $\alpha_c(\log[L])$. (*b*) Percentage plots $\sum_0^c \alpha_c(\log[L])$. (*c*) Log–log plots $\log \alpha_c(\log[L])$. (*d*) Equilibrium ratio diagram $\log[ML_c]/[ML]$ ($\log[L]$).

Fig. 13.4 The system Ag⁺–NH₃ in which $N = 2$, $\log K_1 = 3\cdot37$ and $\log K_2 = 3\cdot78$. (*a*) Distribution plots $\alpha_c(\log[L])$. (*b*) Percentage plots $\sum_0^c \alpha_c(\log[L])$. (*c*) Log–log plots $\log \alpha_c(\log[L])$. (*d*) Equilibrium ratio diagram $\log[ML_c]/[ML]$ ($\log[L]$).

Similarly, from equations (13.3), (13.14) and (13.15), we have

$$\bar{n} = \frac{\rho\mathbf{L} + 2\mathbf{L}^2}{1 + \rho\mathbf{L} + \mathbf{L}^2} \tag{13.19}$$

It is clear that the shapes of plots of α_c, $\log \alpha_c$ or $\bar{n}$ against $\log \mathbf{L}$ depend on the value of the parameter ρ. The shape of any particular plot is the same whether $\log \mathbf{L}$ or $\log[\mathrm{L}]$ is used as ordinate; the pairs of curves are merely separated by a horizontal displacement corresponding to $\frac{1}{2}\log\beta_2$. Since any plot of $\log R_{c,r}$ against $\log[\mathrm{L}]$ is a straight line, its shape is not affected by the value of ρ, although this parameter does determine the relative positions of the lines for a particular system. Taking the intermediate species ML as reference, we may write

$$R_{0,1} = \frac{[\mathrm{M}]}{[\mathrm{ML}]} = \frac{1}{\beta_1[\mathrm{L}]} \tag{13.20}$$

$$R_{1,1} = 1 \tag{13.21}$$

$$R_{2,1} = \frac{[\mathrm{ML}_2]}{[\mathrm{ML}]} = K_2[\mathrm{L}] \tag{13.22}$$

Thus

$$\log R_{0,1} = -\log\beta_1 - \log[\mathrm{L}] = -\log\rho - \log\mathbf{L} \tag{13.23}$$

$$\log R_{1,1} = 0 \tag{13.24}$$

$$\log R_{2,1} = \log\beta_2 - \log\beta_1 + \log[\mathrm{L}] = -\log\rho + \log\mathbf{L} \tag{13.25}$$

The effect of the value of ρ on the different types of diagram is illustrated in Figs 13.2–13.4.

For values of $\rho \geqslant 10^2$ ($K_1 \gg K_2$) complex formation takes place in two regions of [L] which are so far apart that they may be treated as if they were almost independent (Fig. 13.2). The shapes of the plots of α_c, $\sum \alpha_c$ and $\log \alpha_c$ against $\log[\mathrm{L}]$ are then very similar to the analogous curves for the formation of a single complex (cf. Fig. 13.1). The same is true for the formation curve in the two regions $0 \leqslant \bar{n} \leqslant 1$ and $1 \leqslant \bar{n} \leqslant 2$. The value of ρ merely determines the difference in $\log[\mathrm{L}]$ between the two steps. If complex formation occurs in totally separate steps, we have $\alpha_0 = \alpha_1 = 0\cdot5 = \bar{n}$, when $\log[\mathrm{L}] = -\log\beta_1$ (or $\log\mathbf{L} = -\log\rho$) just as for systems in which $N = 1$. Moreover, when $\log[\mathrm{L}] = -\log K_2$ (or $\log\mathbf{L} = \log\rho$) we have $\alpha_1 = \alpha_2 = 0\cdot5$ and $\bar{n} = 1\cdot5$. In the same way, the value of ρ governs the range of free ligand concentration over which the horizontal line $\log R_{1,1} = 0$ lies above the $\log R_{c,1}$ lines for M or ML_2.

As the values of K_1 and K_2 become closer together, the value of ρ decreases; the various types of plot all become compressed along the $\log[\mathrm{L}]$ axis, as complex formation becomes complete over an increasingly narrow range of free ligand concentration (Figs 13.3 and 13.4). The value of α_1 passes through a maximum at a value increasingly below unity; the plot of $(\alpha_0 + \alpha_1)$ against $\log[\mathrm{L}]$ approaches that for α_0. The log–log plots become steeper, and closer together for the three species. The formation curve becomes steeper and the separate steps disappear. Half-integral values of α_c and $\bar{n}$ no longer correspond to values of $\log[L] = -\log K_c$. The equilibrium ratio plots of $\log R_{0,1}$ and $\log R_{2,1}$ become closer together, so that the 'well' in the diagram narrows. When $\rho < 1$, they intersect above the line $\log R_{1,1} = 0$, indicating that ML is never the predominant species for systems in which $K_2 > K_1$.

It may readily be shown by differentiation of equation (13.17) that the value of α_1 is maximal when $L = 1$ (or $[L] = \beta_2^{-\frac{1}{2}}$). At this point we have

$$\alpha_{1(\text{max})} = \frac{(K_1/K_2)^{\frac{1}{2}}}{2 + (K_1/K_2)^{\frac{1}{2}}} \tag{13.26}$$

'(see Fig. 13.5). The slope of the formation curve may be similarly obtained by differentiating equation (13.19), and is found to have a mid-point value of

$$\left(\frac{d\bar{n}}{d\log L}\right)_{\bar{n}=1} = \frac{2}{2 + (K_1/K_2)^{\frac{1}{2}}}$$
$$= 1 - \alpha_{1(\text{max})} \tag{13.27}$$

Regardless of the value of the shape-parameter ρ, all the diagrams discussed above for systems in which $N = 2$ are in some way symmetrical about

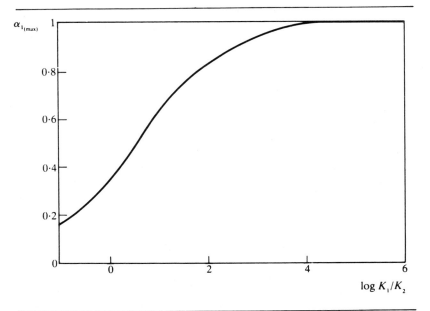

Fig. 13.5 The maximum value of α_1, in a system for which $N = 2$, as a function of $\log K_1/K_2$.

the point where $\log \mathbf{L} = 0$. For distribution diagrams, log–log plots, and equilibrium ratio plots (referred to ML), the symmetry is reflective about the vertical line $\log \mathbf{L} = 0$. Formation curves show rotational symmetry about the mid-point ($\bar{n} = 1$, $\log [L] = -\frac{1}{2} \log \beta_2$) and percentage diagrams show similar symmetry about the point $\alpha_c = 0.5$, $\log [L] = -\frac{1}{2} \log \beta_2$.

'Several' complexes

Systems in which $N \geqslant 3$ may be represented by plots similar to those discussed above but the symmetry of the diagrams for simpler systems is no longer retained (see Fig. 13.6).

We may again replace [L] by a normalised variable, such as

$$[\mathbf{L}] = \beta_N^{1/N}[L] \tag{13.28}$$

The expressions

$$\alpha_c = \frac{\rho_c[\mathbf{L}]}{\sum\limits_0^N \rho_n[\mathbf{L}]^n} \tag{13.29}$$

and

$$\bar{n} = \sum\limits_0^N n\alpha_n \tag{13.30}$$

then contain $(N-1)$ independent parameters $\rho_n = \beta_n \beta_N^{-n/N}$ (since $\rho_N = 1$). So the value of β_N again fixes the positions of the various plots on the $\log [L]$ axis, but since both α_c and $\bar{n}$ contain at least two parameters ρ_n, representing ratios of constants, the shapes of the plots of α_c, $\sum \alpha_c$, $\log \alpha_c$ and $\bar{n}$ against $\log [L]$ are likely to be different for each system. As in systems for which $N = 2$, high values of K_n/K_{n+1} ($\geqslant 10^4$) lead to complete formation of the complex ML_n at free ligand concentrations where ML_{n+1} can be neglected. The value of α_n then reaches a maximum of unity and the formation curve shows a plateau at $\bar{n} = n$.

Lower values of K_n/K_{n+1} imply that the complexes ML_n and ML_{n+1} can coexist at appreciable concentrations. Plots at appropriate values of [L] resemble those for systems in which $N = 2$ and $K_1 \sim K_2$ (cf. Figs 13.3 and 13.4). The maximum value of α_n fails to reach unity and the formation curve in the region where $\bar{n} \sim n$ is steeper than that for a system in which $N = 1$.

The lack of symmetry which arises from the presence of more than one shape-parameter prevents our taking short-cuts of rotation and reflection when constructing the diagrams. More importantly, it also complicates the calculation of stability constants from the original data $\bar{n}$, [L] or α_c, [L]. Systems in which $N = 1$ or $N = 2$ may often be analysed by methods which depend on symmetry (see sect. 4.2) and so clearly cannot be used when three or more complexes are formed.

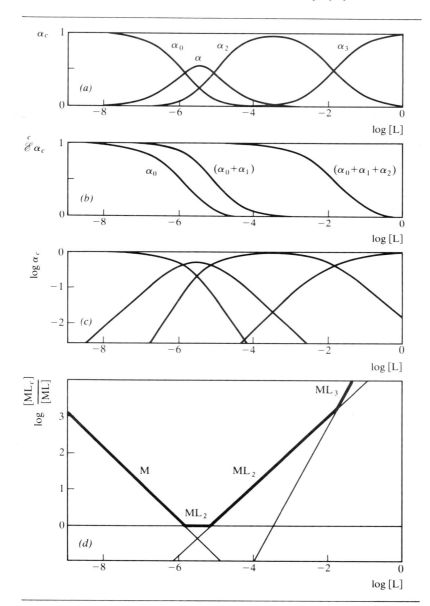

Fig. 13.6 The system Zn^{2+}–ethylenediamine, in which $N = 3$, $\log K_1 = 5 \cdot 92$, $\log K_2 = 5 \cdot 15$ and $\log K_3 = 1 \cdot 86$. (*a*) Distribution plots $\alpha_c(\log[L])$. (*b*) Percentage plots $\sum_0^c \alpha_c(\log[L])$. (*c*) Log–log plots $\log \alpha_c(\log[L])$. (*d*) Equilibrium ratio diagrams $\log[ML_c]/[ML]$, $\log[L]$.

Competitive complex formation

When two sets of binary mononuclear complexes are formed, the system may, of course, be represented by two sets of distribution plots. However, we often need to know how the distribution of one set of complexes depends on the concentration of some other species, in particular how the formation of a set of metal complexes varies with pH.

We may first calculate the functions $\alpha_n(\log[L])$ and $\alpha_j^H(\log[H])$ for the metal complexes ML_n and the acids H_jL. Since $\sum c\alpha_c$ gives $\bar{n}$, we may obtain the value of

$$\frac{L_t - \bar{n}M_t}{[L]} = \beta_j^H[H]^j = (\alpha_0^H)^{-1} \tag{13.31}$$

for any pair of values of L_t and M_t (cf. equations (7.23) and (7.24)). The value of $[H]$ corresponding to this value of α_0^H may then be obtained from the distribution plot for the acid system. Although the functions $\alpha_n(\log[H])$ so obtained are valid only for particular values of the total concentrations of metal ion and ligand, they give a practical representation of how a particular system behaves (see Fig. 13.7).

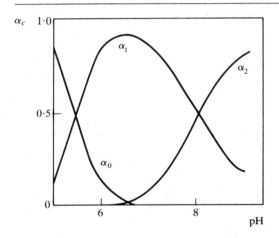

Fig. 13.7   Values of α_c as a function of pH for Co^{2+}–histidine complexes in solutions in which $M_t = 5 \times 10^{-4}M$ and $L_t = 1 \cdot 5 \times 10^{-3}M$. (From M. T. Beck, *Chemistry of Complex Equilibria*, Van Nostrand Reinhold, London (1970).)

Mixed and polynuclear complexes

The distribution of mononuclear complexes $ML_nL_{n'}$ in ternary systems depends on the free concentration of both types of ligand and the formation of binary polynuclear complexes M_qL_p similarly depends on the two variables $[M]$ and $[L]$. Diagrams of the types discussed above can therefore only be used to represent ternary, or polynuclear, systems if one of the variables is held constant.

In order to represent the formation of mixed mononuclear complexes, we might construct two sets of equilibrium ratio diagrams, one being plots of $\log R_{nn',00}$ against $\log [L]$ at a number of fixed values of $[L']$ and the other being a similar relationship $\log R_{nn',00}(\log [L'])_{[L]}$ with fixed $[L]$ but variable $[L']$. The uncomplexed metal ion should be chosen as the reference species, since it is the only species for which the concentration can be calculated both when $[L] = 0$ and when $[L'] = 0$. A full representation would require three dimensions, with the axes representing $\log R_{nn',00}$, $\log [L]$ and $\log [L']$. A relief model, with each species represented by a coloured, transparent hummock, would be the best way to display the system, given adequate resources for making one. A section through such a model, at a constant value of $\log [L']$, would generate the equilibrium ratio diagram $\log R_{nn',00}(\log [L])_{[L']}$.

Let us imagine that the two horizontal axes of the three-dimensional model represent $\log [L]$ and $\log [L']$. The vertical axis then gives $\log R_{nn',00}$ and the uppermost surface of the model will show only those species which

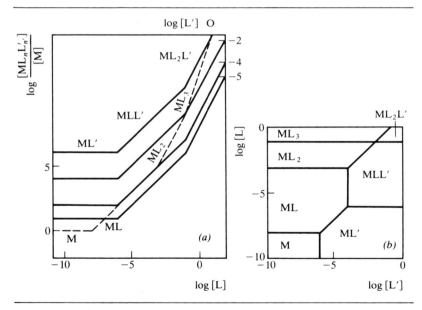

Fig. 13.8 The hypothetical ternary system M, L, L' in which the following species are present:

ML	$\log \beta_{1,0} = 8$	ML'	$\log \beta_{0,1} = 6$
ML$_2$	$\log \beta_{2,0} = 11$	MLL'	$\log \beta_{1,1} = 12$
ML$_3$	$\log \beta_{3,0} = 12$	ML$_2$L'	$\log \beta_{2,1} = 13$

(*a*) Equilibrium ratio plots, relative to M. Dotted line, $[L'] = 0$. Full lines from left to right, $\log [L'] = 0, -2, -4, -5$. (*b*) Predominance-area diagram.

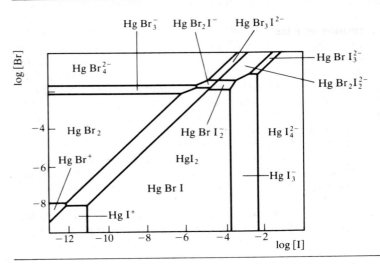

Fig. 13.9 Predominance-area diagram for the mixed mononuclear complexes of Hg^{2+} with Br^- and I^-. (After J. N. Butler, *Ionic Equilibria*, Addison-Wesley, Reading, Mass. (1964).)

predominate at some pair of values of [L] and [L']. We can draw a projection of this upper surface, as a plot of log [L] against log [L'], to map out the areas of ligand concentrations in which each of these species predominates. An example of such a *predominance-area diagram* is shown in Fig. 13.8, constructed for the hypothetical system M, L, L' with stability constants shown in the caption. The system was first represented by drawing a set of equilibrium ratio diagrams $\log R_{nn',00}(\log [L])_{[L']}$ for various values of log [L'], in order to see which of the species were, in fact, predominant. It emerges that every possible species is predominant at some combination of [L] and [L'] and so must be considered. The predominance-area diagram is drawn by considering pairs of adjacent species. The equation for the boundary between the two predominance areas is found by equating the concentrations of the appropriate species. Thus, for the boundary between M and ML we have

$[M] = [ML]$ whence $\log [L] = -\log \beta_{10}$. Similarly, when

$[M] = [ML']$ $\log [L'] = -\log \beta_{0,1}$

$[ML] = [ML']$ $\log [L] = \log [L'] + \log \beta_{0,1} - \log \beta_{1,0}$

$[ML] = [MLL']$ $\log [L'] = \log \beta_{1,0} - \log \beta_{1,1}$

and so on.

The general equation of the boundary separating the two predominant species $ML_{n_1}L'_{n'_1}$ and $ML_{n_2}L'_{n'_2}$ is

$$(n_1 - n_2) \log [L] = (n'_2 - n'_1) \log [L'] + \log \beta_{n_2 n'_2} - \log \beta_{n_1 n'_2} \qquad (13.32)$$

Boundaries between areas for species with the same number of one type of ligand are therefore parallel to the appropriate axis (see Figs 13.8 and 13.9).

The formation of mixed complexes by replacement reactions in which both the coordination number and the sum of the concentrations of ligands are unchanged (see sect. 11.2) may be represented in two dimensions as distribution plots of $\alpha_{cc'} = ML_cL'_{c'}/M_t$ against $\log L_t/L'_t$. A plot of this type, for the species BiI_3, BiI_2Cl, $BiICl_2$ and $BiCl_3$, is shown in Fig. 13.10.

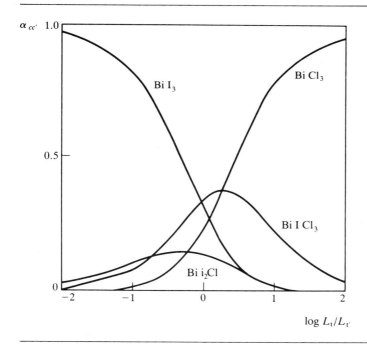

Fig. 13.10 Distribution plots for the system Bi^{3+}, Cl^-, I^-. Values of $\alpha_{cc'}$ for complexes in which $(c + c') = 3$ as a function of $\log Cl_t/I_t$. (From M. T. Beck, *Chemistry of Complex Equilibria*, Van Nostrand Reinhold, London (1970).)

Binary polynuclear systems may be treated in exactly the same way as mixed complexes $ML_nL_{n'}$. Equilibrium ratio diagrams $\log R_{qp,r}(\log [L])_{[M]}$ may be constructed for a series of values of $[M]$ in order to identify the predominant species. Any convenient species may be chosen as reference. A predominance-area plot of $\log [L]$ against $\log [M]$ may then be drawn (see Figs 13.11 and 13.12).

Although predominance-area diagrams take no account of any minority species, they are nonetheless extremely useful for providing an overall view of the system.

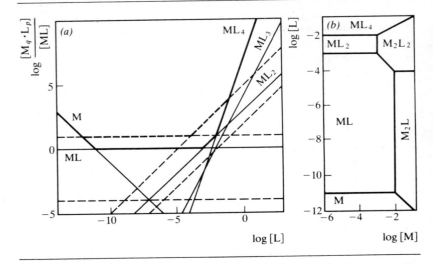

Fig. 13.11 The hypothetical polynuclear system M, L in which the following species are present:

ML $\log \beta_{1,1} = 11$ ML$_4$ $\log \beta_{1,4} = 18$
ML$_2$ $\log \beta_{1,2} = 14$ M$_2$L $\log \beta_{2,1} = 13$ (horizontal dashed lines)
ML$_3$ $\log \beta_{1,3} = 15$ M$_2$L$_2$ $\log \beta_{2,2} = 17$ (oblique dashed lines)

(*a*) Equilibrium ratio plots, relative to ML, for values of $\log [M] = 10^{-1}$ M (upper dashed lines) and 10^{-4} M (lower dashed lines). (*b*) Predominance-area diagram.

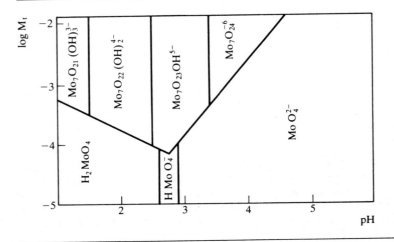

Fig. 13.12 Predominance-area plot for the polynuclear hydrolysis of molybdenum(VI). (From C. F. Baes and R. E. Mesmer, *The Hydrolysis of Cations*, John Wiley, New York (1976).)

13.3 Treatment of data M_t, L_t

We often need to know the equilibrium concentrations in a solution for which only the overall composition is known. The problem is not too difficult when a single series of binary mononuclear complexes, such as ML_n, is formed. The term [M] may be eliminated from the mass balance equations

$$M_t = \sum [ML_n] = [M] \sum \beta_n [L]^n \tag{13.33}$$

and

$$L_t = [L] + \sum n[ML_n] = [L] + [M] \sum n\beta_n [L]^n \tag{13.34}$$

giving a polynomial which may then be solved for [L]. Even if computational facilities are not available, the value of [L] may often be obtained without undue hardship by either numerical or graphical methods. Once [L] is known, the value of $[ML_c] = \alpha_c M_t$ can be calculated via equation (7.24).

A convenient numerical procedure involves guessing a value of [L], using it to calculate a value of $\bar{n}$, and refining the pair of values $\bar{n}$, [L] by successive approximations. First, we may assume

$$[L_1] \sim L_t \tag{13.35}$$

and calculate

$$\bar{n}_1 = \sum n\beta_n [L_1] / \sum \beta_n [L_1]^n \tag{13.36}$$

Then we can use

$$[L_2] = L_t - \bar{n}_1 M_t \tag{13.37}$$

to calculate a better value, $\bar{n}_2$, of $\bar{n}$; and so on, until successive values of $\bar{n}$, [L] are identical. A similar iterative procedure, involving Newton's method, forms the basis of the solution of equation (13.36) by computer.

The value of [L] in a solution of known L_t and M_t may also be obtained graphically. The formation curve $\bar{n}(\log [L])$ is calculated for a range of free ligand concentrations, using equation (13.36). Values of

$$\bar{n}' = (L_t - [L']) / M_t \tag{13.38}$$

are also calculated using the appropriate pair of values L_t, M_t and a number of values of [L']. The point of intersection of the function $\bar{n}'(\log [L'])$ with the formation curve gives the values of $\bar{n}$ and [L] for a solution of composition M_t and L_t (see Fig. 13.13).

If we need to solve equation (13.34) frequently for a particular system, we may construct an alignment chart, or nomogram, such as that devised by Butler for cadmium chloride complexes (see Fig. 13.14). A straight line through the appropriate points on the two vertical scales, representing L_t and M_t, cuts the curved scale at the required value of [L]. The equations for the

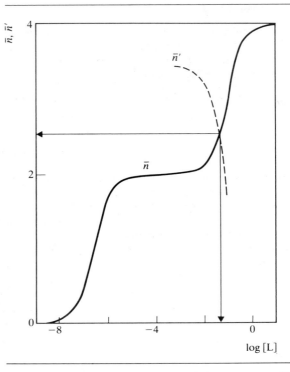

Fig. 13.13   Graphical determination of [L] for the system Hg^{2+}–Cl^-. The dashed line is the function $\bar{n}'$, log [L] calculated using equation (13.38) with $M_t = 0.05$M and $L_t = 0.18$M. (From W. B. Guenther, *Chemical Equilibrium*, Plenum, New York (1975).)

three scales are, from left to right,

L_t scale:

$$y = mL_t \qquad x = 0 \tag{13.39}$$

[L] scale:

$$y = \frac{\bar{n} + m[L]}{1 + \bar{n}} \qquad x = \frac{\bar{n}}{1 + \bar{n}} \tag{13.40}$$

M_t scale:

$$y = 1 - mM_t \qquad x = 1 \tag{13.41}$$

The value of m, which must be the same for all three scales, is chosen so that the chart covers a convenient range of concentrations. The pairs of values of $\bar{n}$, [L] needed for the [L] scale are calculated using equation (13.36). The equations (13.39) to (13.41) can be used for any binary mononuclear complexes ML_n provided that neither metal ion nor ligand take part in protonic equilibria.

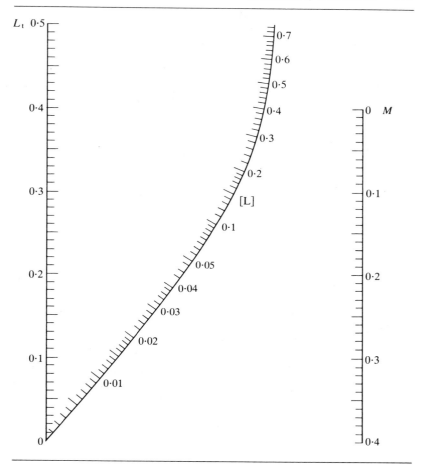

Fig. 13.14 Nomogram for the Cd^{2+}–Cl^- system. (From J. N. Butler, *Ionic Equilibria*, Addison-Wesley, Reading, Mass. (1964).)

Once designed, an alignment chart of this type is easy to draw and extremely convenient to use; to find [L] for a solution of known overall composition we need only place any straight edge across the two outer scales. If the precision is not high enough, the value of [L] may be refined iteratively, as described on p. 169.

13.4 Treatment of data H_t, L_t

The problem of calculating [H] in solutions containing acids H_jL is, of course, formally analogous to that of obtaining the value of [L] for a solution containing the complexes ML_n. But there are two practical differences. One, which operates to our advantage, is that, because stepwise formation constants of acids tend to be much more widely spaced than those of metal complexes we need often consider only one protonation step at given pH. The second, but disadvantageous, difference is that we cannot treat L, and its acids, as the only species which take part in protonic equilibria in aqueous solution. The situation is complicated by the dissociation of water itself and although the products of this dissociation can often be neglected, they should never be forgotten.

The total concentration of dissociable protons is therefore not exactly analogous to L_t, for metal–ligand systems, but is given by

$$H_t = [H] - [OH] + \sum_0^J j[H_jL]$$

$$= [H] - K_w[H]^{-1} + [L] \sum_0^J \beta_j^H [H]^j \qquad (13.42)$$

Since the total concentration of L is

$$L_t = \sum_0^J [H_jL] = [L] \sum_0^J \beta_j^H [H]^j \qquad (13.43)$$

the value of [L] can be eliminated from equations (13.42) and (13.43) and the resulting polynomial solved for [H] provided that values of the $(J+1)$ parameters, $\beta_1^H \ldots \beta_J^H$ and K_w, are known. We may use either numerical or graphical methods, analogous to those described in the previous section.

For solving particular problems, it may be convenient to involve an expression for the balance of either transferred protons or electric charge instead of one of the more general mass balance equations (13.42) or (13.43). For example, if NaH_2PO_4 is added to water, the total number of protons lost from the $H_2PO_4^-$ ion and from water must equal that gained by other species. This *proton balance* can be expressed as

$$[HPO_4^{2-}] + 2[PO_4^{3-}] + [OH^-] = [H_3PO_4] + [H_3O^+] \qquad (13.44)$$

proton-deficient species   proton-rich species
(relative to $H_2PO_4^-$ (relative to
and H_2O) $H_2PO_4^-$ and H_2O)

We may alternatively use an expression for *charge balance*. Since the solution must be electrically neutral, the total concentration of positive and negative charges must be equal. For an aqueous solution of NaH_2PO_4, we therefore have

$$[Na^+] + [H_3O^+] = [H_2PO_4^-] + 2[HPO_4^{2-}] + 3[PO_4^{3+}] + [OH^-] \qquad (13.45)$$

For monobasic acids, and for polybasic acids with widely separated pK values, only little practice is needed in deciding which species we can properly ignore. It is obvious that the hydroxyl ion concentration decreases with decreasing pH: and the stronger the acid, and the higher its concentration, the less important [OH] becomes. A convenient approach to the calculation of pH is to make reasonable assumptions about which species to neglect, obtain a preliminary value for [H], and use this value to check whether or not the simplifications were justified. If they were not, then the calculations must be repeated using equations which take fuller account of the equilibria involved. The refined calculation will require the solution of a higher-order polynomial; but this can be performed by Newton's method, using the preliminary value of [H] as the approximate root needed as starting-point.

It may be more convenient to use log–log plots to ascertain which species must be considered, particularly if a mixture of acids is present. Values of $\log[HL]$ and $\log[L]$ are plotted against $\log[H]$ for each acid–base pair. The diagram may readily be obtained from the normalised plots of $\log \alpha_1$ and $\log \alpha_0$ against $\log h$ (Fig. 13.1(c)) if the origin of the normalised curves is displaced to the point $\log L_t$, $-\log \beta_1^H$ for the particular solution of acid. The straight lines representing $\log[H]$ and $\log[OH]$ are then superimposed.

The applications of log–log plots have been described in detail by Butler and only a few examples will be outlined here. Figure 13.15 represents a 0·01M solution of acetic acid in water. From the electro-neutrality relationship, we have

$$[H^+] = [L^-] + [OH^-] \tag{13.46}$$

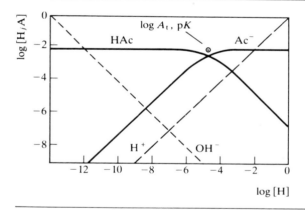

Fig. 13.15 Log–log concentration diagram for 0·01M acetic acid ($pK = 4·74$). (After J. N. Butler, *Ionic Equilibria*, Addison-Wesley, Reading, Mass. (1964).)

and it is clear from the diagram that this condition is fulfilled at a point where $[H^+] = [L^-]$ and $[OH^-]$ is negligible. Thus pH = 3·37.

A similar plot for 10^{-4}M hydrocyanic acid is shown in Fig. 13.16. Since this solution, too, is electrically neutral, we know that

$$[H^+] = [CN^-] + [OH^-] \tag{13.47}$$

The diagram shows that the lines for $[CN^-]$ and $[OH^-]$ lie quite close together; at no pH does $[OH^-]$ form less than 10 per cent of total

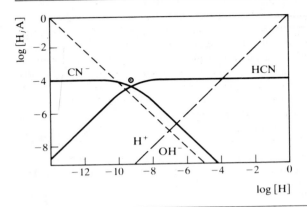

Fig. 13.16   Log–log concentration diagram for 10^{-4}M hydrocyanic acid (p$K = 9.32$). (After J. N. Butler, *Ionic Equilibria*, Addison-Wesley, Reading, Mass. (1964).)

concentration of anions. So we can obtain only an approximate value of [H] from the point where the lines for log [H] and log [CN⁻] intersect. The definitive value of [H] must then be obtained by successive approximations using equation (13.40). Figure 13.16 again guides us by showing that [HCN] ≫ [CN⁻] in the regions where [H⁺] = [CN⁻] and [H⁺] = [OH⁻]. We may therefore use the simplification [HCN] ≃ L_t.

Figures 13.17 and 13.18(*a*) show how the addition of an acid HL′ affects the pH of a 0·1M aqueous solution of acetic acid, HL. The pH is virtually unchanged by the addition of a high concentration of a much weaker acid, or by a very low concentration of a strong acid. Thus both [L′⁻] and [OH⁻] are

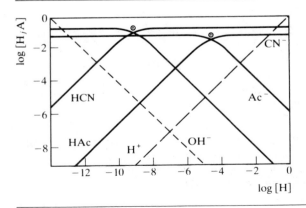

Fig. 13.17   Log–log concentration diagram for a solution containing 0·1M acetic acid (p$K = 4.74$) and 0·25M hydrocyanic acid (p$K = 9.32$). (After J. N. Butler, *Ionic Equilibria*, Addison-Wesley, Reading, Mass. (1964).)

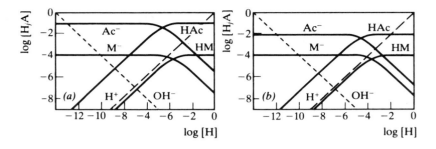

Fig. 13.18 Log–log concentration diagrams for solutions containing acetic acid (pK = 4·74) and 10^{-4}M methyl orange (pK = 3·8), HM. (*a*) 0·1M acetic acid. (*b*) 0·01M acetic acid. (After J. N. Butler, *Ionic Equilibria*, Addison-Wesley, Reading, Mass. (1964).)

negligible compared with [L$^-$] for 0·25M HCN (pK = 9·32, Fig. 13.17) and 10^{-4}M methyl orange (pK = 3·8, Fig. 13.18(*a*)), so that we may set [H$^+$] = [L$^-$] in both these mixtures, as in pure 0·1M acetic acid. However, vertical displacement of the acetic acid plots to log L_t = −2 shows (Fig. 13.18(*b*)) that addition of 10^{-4}M methyl orange to 0·01M acetic acid would cause an appreciable increase in [H$^+$], since [L$^-$] and [L$'^-$] would be of comparable magnitude. Similarly, addition of an acid HL to a solution of concentration comparable with that of HL would cause an appreciable increase in [H$^+$] if pK_{HL} is less than 2 units greater than pK_{HL} (see Fig. 13.19).

The application of log–log plots to polyprotic equilibria may be illustrated by Fig. 13.20, which represents the 0·1M phosphate system. If we want to

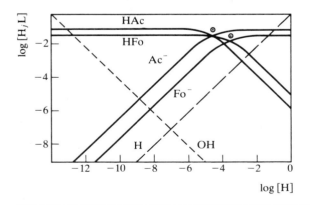

Fig. 13.19 Log–log concentration diagram for a solution containing 0·1M acetic acid (pK = 4·74) and 0·05M formic acid, HFo (pK = 3·75). (After J. N. Butler, *Ionic Equilibria*, Addison-Wesley, Reading, Mass. (1964).)

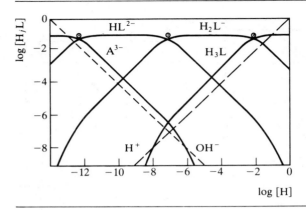

Fig. 13.20 Log–log concentration diagram for $0 \cdot 1M$ phosphoric acid, H_3PO_4 (pK values: $2 \cdot 23$, $7 \cdot 21$ and $12 \cdot 32$). (After J. N. Butler, *Ionic Equilibria*, Addison-Wesley, Reading, Mass. (1964).)

know the pH of a $0 \cdot 1M$ solution of NaH_2PO_4, the proton balance condition, relative to $(H_2L + H_2O)$, is

$$[H_3L] + [H] = [HL] + 2[L] + [OH] \tag{13.48}$$

Moreover, between pH $\sim 3 \cdot 5$ and pH $\sim 6 \cdot 5$, we have $\alpha_2 \sim 1$ so that $[H_2L] \simeq L_t$. The diagram shows that $[L]$ and $[OH]$ may be neglected in this range and that $[H]$ is small, but not negligible, compared with $[H_3L]$. A good approximation to $[H]$ may be obtained from the point where $[H_3L] = [HL]$, and refined using equation (13.42).

The value of $[H]$ in $0 \cdot 1M$ Na_2HPO_4 may be obtained similarly from the range $8 \cdot 5 < \text{pH} < 11 \cdot 5$, where $\alpha_1 = 1$ and $[HL] \simeq L_t$. Then, from the proton balance relationship, relative to $(HL + H_2O)$,

$$2[H_3L] + [H_2L] + [H] = [L] + [OH] \tag{13.49}$$

The diagram shows that only the species H_2L, L and OH need be considered. If $[OH]$ is (unjustifiably) neglected, the value of pH $\sim 9 \cdot 75$ is obtained as the point where $[H_2L] = [L]$; numerical calculation gives pH $= 9 \cdot 72$. So for rough work, even the original estimate might be acceptable.

Figure 13.21 may also be used to estimate the pH of $0 \cdot 1M$ $(NH_4)_2HPO_4$ if we add the curves for ammonia, L', and the ammonium ion HL' in the position for $L'_t = 0.2M$. We can express the proton balance, relative to the initial species $(NH_4^+, HPO_4^{2-}$ and $H_2O)$ as

$$2[H_3L] + [H_2L] + [H] = [L'] + [L] + [OH] \tag{13.50}$$

Inspection of the diagram shows that this condition is fulfilled at pH $= 8 \cdot 07$, where $\alpha_{1(L)} = 1$, $\alpha_{1(L')} = 1$, $[H_2L] = [L]$ and all other species may be neglected.

This brief outline does scant justice to the range of problems which may

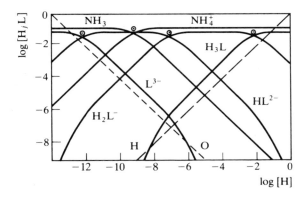

Fig. 13.21 Log–log concentration diagram for $0 \cdot 1 M$ $(NH_4)_2HPO_4$ solution (pK for $NH_4^+ = 9 \cdot 25$). (After J. N. Butler, *Ionic Equilibria*, Addison-Wesley, Reading, Mass, (1964).)

be solved with the help of log–log plots. The diagrams are extremely easy to construct, and fun to use, particularly if a set of different coloured normalised curves is prepared on separate transparent sheets. They may then be superimposed on a backing sheet marked with log L_t and log [H] scales, and showing log concentration plots for H and OH. (In polyprotic systems, regions in which the slope is steeper than ± 1 may usually be ignored.) The answers given by log–log diagrams are, of course, no better than those obtained by calculation; indeed they are often slightly worse. But they are obtained using negligible time and effort, and give invaluable information about which species may, or may not, be neglected. Even when the values need refining, they are so near to the convergence value that little further work is needed.

When a number of complexes coexist, as in many solutions of metal complexes, fewer species may be neglected and so log–log plots are much less useful.

13.5 Polynuclear and ternary systems

Calculation of equilibrium concentrations in systems which contain polynuclear or mixed complexes, or even in binary systems which contain two series of complexes (e.g. $ML_n + H_jL$), is much more tiresome than the examples discussed in the two previous sections. These more complex systems are best handled by numerical solution of the mass balance expressions for M_t, L_t and H_t, unless grossly simplifying assumptions can be made in certain concentration ranges. As the calculations often involve solution of polynomials of a high order, they are best performed by a computer.

3.6   Heterogeneous equilibria

Methods similar to those described above may be used to calculate concentrations of species in equilibrium with a solid, or second liquid, phase. In addition to the stability constants, we merely need to know the value of one distribution parameter for each species which partitions. For solubility studies, we normally use the solubility product K_{so}, which is the equilibrium constant for the reaction

$$ML_c(s) \rightleftharpoons M(aq) + c\,L(aq) \tag{13.51}$$

but if, like $HgCl_2$, the solid dissolves mainly as the uncharged complex, we may instead use its intrinsic solubility, K_{sc}, which is the equilibrium constant for the partition process

$$ML_c(s) \rightleftharpoons ML_c(aq) \tag{13.52}$$

Clearly,

$$K_{sc} = \beta_c K_{so} \tag{13.53}$$

Studies of the distribution of an uncharged species between two liquid phases make use of its partition coefficient, P_c, defined as the equilibrium constant for the reaction

$$ML_c(o) \rightleftharpoons ML_c(aq) \tag{13.54}$$

We often need to predict, or interpret, the solubility S of ML_c in one of a number of aqueous media, such as:
  water
  a solution of NaL (of concentration L_i)
  a solution of a strong acid (of concentration H_t)
  a buffer solution
  a solution containing a ligand L' (of concentration L'_t)
We again express the total concentration of M, L, L' and H in terms of the component species, remembering (cf. equation (9.4)) that

$$M_t = S = [M] \sum \beta_n [L]^n = K_{so} \sum \beta_n [L]^{n-c} \tag{13.55}$$

and

$$L_t = cS + L_i = [M] \sum n\beta_n [L]^n = K_{so} \sum n\beta_n [L]^{n-c} \tag{13.56}$$

As with homogeneous equilibria, the calculations are much easier if the equilibrium concentration of one species is known. For example, if the solubility of ML_c in solutions of NaL is very low for all concentrations, we may assume $L_t \simeq L_i \simeq [L]$. The concentrations of the complexes may then easily be calculated as

$$[ML_n] = K_{so}\beta_n [L]^{n-c} \tag{13.57}$$

or

$$[ML_n] = K_{sc}\beta_n\beta_c^{-1}[L]^{n-c} \tag{13.58}$$

Similarly, if the aqueous solution is a buffer of known [H], we can calculate

$$\alpha_0^H = (\sum \beta_j^H[H]^j)^{-1} \tag{13.59}$$

Now from equation (13.29)

$$[L] = \alpha_0^H(L_t - \bar{n}M_t) \tag{13.60}$$

and, when $L_i = 0$, we have

$$[L] = \alpha_0^H S(c - \bar{n}) \tag{13.61}$$

Moreover, if $\bar{n} = 0$, and ML_c dissolves effectively as M, L and the acids H_jL, we may write

$$S = [M] = K_{so}[L]^{-c} \tag{13.62}$$

and

$$[L]^{c+1} \simeq \alpha_0^H K_{so}c \tag{13.63}$$

When the value of [H] is not known, it may be found by solving the appropriate mass balance expression. For example, the solubility of a metal fluoride MF_c in nitric acid is given by

$$S = M_t = \sum [MF_n] = K_{so} \sum \beta_n[F]^{n-c} \tag{13.64}$$

We combine equation (13.62) with the expressions

$$F_t = cS = [F] + [HF] + \sum n[MF_n]$$
$$= [F]\{1 + \beta_1^H[H] + K_{so} \sum n\beta_n[F]^{n-c-1}\} \tag{13.65}$$

and

$$H_t = [H] + [HF] - [OH] = [H]\{1 + \beta_1^H[F] - K_w/[H]^2\} \tag{13.66}$$

Since H_t is known, equations (13.62) to (13.66) may be solved for [H], [F] and S.

Calculations of solubility as a function of L_i may often be simplified by the justifiable neglect of certain species; and, as with homogeneous systems, log concentration diagrams may be useful for deciding which approximations to make. The log concentration diagram for the Ag^+–SCN^- system, including the line for log [SCN^-], is shown in Fig. 13.22. The solubility at $L_i = 0$ can be found using the electro-neutrality relationship.

$$[Ag^+] = [SCN^-] + [AgSCN^-] + 2[Ag(SCN)_2^{2-}] \tag{13.67}$$

The diagram shows that the concentrations of the complexes are negligible at the point where $[Ag^+] = [SCN^-]$, so that $S = K_{so}^{\frac{1}{2}}$. In 0·1M NaSCN on the other hand, the silver ion is almost entirely in the form of either $Ag(SCN)_3^{2-}$ or $Ag(SCN)_4^{3-}$, while the value of [SCN^-] is effectively L_i. In this and in more

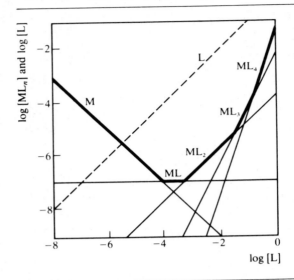

Fig. 13.22 Equilibrium ratio diagram, relative to AgSCN, for the system Ag^+–SCN^-, displayed so that the horizontal line represents the concentration of undissociated AgSCN in contact with its solid: The dotted line represents $[SCN^-]$. (From W. B. Guenther, *Chemical Equilibrium*, Plenum, New York (1975).)

concentrated solutions of NaSCN, the solubility of AgSCN is therefore

$$S = K_{so}L_i^2(\beta_3 + \beta_4 L_i) \tag{13.68}$$

Liquid–liquid partition may be handled in the same way as solubility, but it is slightly more awkward because concentrations now vary in both phases. Since the phases need not be of equal volume, the mass balance expressions must now refer to masses, rather than to concentrations. Thus the total mass of metal ion in the two phases (o) and (w) of volume V_o and V_w is given by

$$M_m = M_{m,o} + M_{m,w}$$
$$= V_o M_{t,o} + V_w M_{t,w} \tag{13.69}$$

If all the metal ion was originally present in one phase of volume V_i and concentration $M_{t,i}$, we may write

$$V_i M_{t,i} = V_o M_{t,o} + V_w M_{t,w} \tag{13.70}$$

which may be simplified to

$$M_{t,i} = M_{t,o} + M_{t,w} \tag{13.71}$$

provided that equal phase volumes are used and that no change in volume occurs on equilibration.

From equation (9.14), the distribution ratio of M between an aqueous phase and an organic phase which contains only one complex ML_c is given by

$$q = P_c\alpha_c \qquad (13.72)$$

The equilibrium state of the system can therefore be calculated with the aid of equations (13.69) to (13.72) provided that we know the total initial concentrations of metal ion, ligand and hydrogen ions, the initial and final phase volumes, the partition coefficients P of all uncharged species (including ion-pairs) and, of course, the stability constants.

13.7 The evaluation of other parameters

We shall now discuss how we can use stability constants to determine a variety of other parameters, including other equilibrium constants (for both homogeneous and heterogeneous reactions), thermodynamic functions and intensive factors.

We are already familiar with various ways in which knowledge of values of some equilibrium constants is necessary for the calculation of others. Thus the stability constants β_n of complexes ML_n of a basic ligand can only be obtained from measurements of, for example, M_t, L_t, H_t and [H] if we know the protonation constants β_j^H for the formation of the acids H_jL from L.

If we know values of two related equilibrium constants, we may often calculate the equilibrium constant of a third reaction. For example the ratio β_2/β_1 of the first two overall stability constants gives the stepwise stability constant K_2 for the formation of ML_2 from ML. The ratio of the formation constants for the species ML and HL similarly gives the equilibrium constant, K_{disp}, for the displacement reaction

$$M + HL \rightleftharpoons ML + H \qquad (13.73)$$

$$K_{disp} = \beta_1/\beta_1^H \qquad (13.74)$$

We have seen that heterogeneous equilibrium constants, such as solubility products and partition coefficients, may be obtained together with the stability constants, from distribution measurements (see sect. 13.6). It may also be possible to measure the stability constants independently and combine them with distribution data to obtain the partition parameter.

Now, the solubility S of a sparingly soluble metal salt ML_c is

$$S = M_t = [M] \sum \beta_n[L]^n = K_{so}\alpha_0^{-1}[L]^{-c} \qquad (13.55)$$

and the distribution ratio of M between two immiscible solvents is given by

$$q = \frac{M_{t,o}}{M_{t,w}} = P_c\alpha_c \qquad (13.75)$$

provided that only the species ML_c partitions. Thus if S or q has been measured as a function of [L], and if the stability constants have been measured independently under identical conditions, we may calculate α_0 or

α_c as a function of [L] and readily obtain the partition parameters as

$$K_{so} = S\alpha_0[L]^c \tag{13.76}$$

and

$$P_c = q\alpha_c^{-1} \tag{13.77}$$

The complex ML_c must, of course, be sufficiently soluble in water for the stability constants to be measured in a one-phase system.

This use of independently measured stability constants may be extended so that it can be applied to the determination of parameters, such as extinction coefficients, which describe the behaviour of several components of a solution. From equations (10.4) and (10.8), the optical absorbency of a solution containing complexes ML_n is given by

$$A_s l^{-1} = \varepsilon_L[L] + \sum_0^N \varepsilon_n[ML_n] \tag{13.78}$$

so that

$$\mathscr{E} = \frac{A_s l^{-1} - \varepsilon_L[L]}{M_t} = \frac{\sum_0^N \varepsilon_n\beta_n[L]^n}{\sum_0^N \beta_n[L]^n} \tag{13.79}$$

$$= \alpha_0 \sum_0^N \varepsilon_n\beta_n[L]^n \tag{13.80}$$

It is possible (see sect. 10.1) to measure $\mathscr{E}$ as a function of [L] and solve equation (13.77) for the $(2N+1)$ parameters ε_n and β_n. But the values of extinction coefficients can be obtained much more easily, and more precisely, if the values of β_n are already known. The values of $\sum \beta_n[L]^n$, and hence also α_0, may be calculated for each value of [L] so that equation (13.77) need now be solved for only $(N+1)$ unknowns, ε_n.

Equations analogous to (13.77) can be set up to describe other properties, such as the observed change in enthalpy on mixing M and L. Combination of values of β_n with the calorimetric data leads to values, ΔH_n, of the enthalpy change for the formation of each of the complexes ML_n. These values, like the extinction coefficients, are independent of the concentrations of the complexes, provided that a large excess of bulk electrolyte is used.

Equations formally similar to (13.80) can also be written to describe other properties such as the electrical conductivity of a solution or the distribution of a metal ion between a cation exchange resin and an aqueous phase (see sect. 9.3). When the stability constants are known, values of ionic mobilities, u, and resin-to-water partition coefficients, P, can be calculated from the variation of conductivity, or distribution, with [L]. It is extremely difficult to obtain reliable values of u and P without introducing independent values of stability constants because, as we have seen, they are not true parameters, but vary appreciably with the composition of the system.

The variation of stability constants with temperature

The temperature-dependence of a stability (a protonation) constant is obtained from the familiar second-law equation

$$\Delta G_n^{\ominus} = \Delta H_n^{\ominus} - T \, \Delta S_n^{\ominus} = -RT \ln \beta_n \tag{13.81}$$

where $\Delta G_n^{\ominus}$, $\Delta H_n^{\ominus}$ and $\Delta S_n^{\ominus}$ are the standard changes in free energy, enthalpy and entropy accompanying the change

$$M + nL \rightarrow ML_n \tag{13.82}$$

for which β_n is the equilibrium quotient at constant (normally atmospheric) pressure, $\mathscr{P}$, and the standard state is infinite dilution of M, L and H in the appropriate ionic medium. It follows from equation (13.81) that

$$\ln \beta_n = -\frac{\Delta H_n^{\ominus}}{RT} + \frac{\Delta S_n^{\ominus}}{R} \tag{13.83}$$

whence

$$\left(\frac{\partial \ln \beta_n}{\partial T}\right)_{\mathscr{P}} = \frac{\Delta H_n^{\ominus}}{RT^2} \tag{13.84}$$

so that, by integration,

$$\ln \frac{\beta_{n'}}{\beta_{n''}} = -\frac{\Delta H_n^{\ominus}}{R}\left(\frac{1}{T'} - \frac{1}{T''}\right) \tag{13.85}$$

where $\beta_{n'}$ and $\beta_{n''}$ are the values of n at temperatures T' and T'' *in a range over which $\Delta H_n^{\ominus}$ is effectively constant.* Equation (13.85) might suggest that we may obtain the value of $\Delta H_n^{\ominus}$ merely by measuring β_n at two temperatures. However, it is obviously preferable to work at several temperatures and obtain $\Delta H_n^{\ominus}$ from the gradient plot of the plot of $\ln \beta_n$ against T^{-1}. It might seem that the use of three or more temperatures would serve as a check on whether or not the italicised condition were fulfilled: if $\Delta H_n^{\ominus}$ varies with temperature, $\ln \beta_n$ will not vary linearly with T^{-1}. Unfortunately, it is often difficult to make this check. Stability constants can usually only be measured over a temperature range of about 30, or at most, 40°C and the variation in $\ln \beta_n$ may not be large enough to show unequivocally that $\ln \beta_n$ is a linear function of T^{-1}, particularly if the values of β_n are not of the highest precision.

Since values of $\Delta H_n^{\ominus}$ for many acids and metal complexes have been obtained by this potentially unreliable method, the thermodynamic functions quoted in the literature should be viewed with initial caution. It is much better to determine enthalpies of complex formation by measuring the heat of reaction, ΔH, calorimetrically and combining it with independent values of the stability constants. For example, if we measure ΔH as a function of L_t and M_t for a system in which the values of β_n have been measured, we can calculate the functions $\Delta H([L])$ and $\alpha_0([L])$ (see sect. 13.3). Since

$$\Delta H = \sum \Delta H_n [ML_n] = \alpha_0 M_t \sum \Delta H_n \beta_n [L]^n \tag{13.86}$$

the values of ΔH_n for each complex may be calculated for the particular temperature of the measurements. The procedure is exactly analogous to that described on p. 182 for obtaining extinction coefficients.

The individual entropy changes of complex formation, ΔS_n, may be calculated by combining the enthalpies of complex formation with the stability constants. From equation (13.81) we have

$$S_n = R \ln \beta_n + \Delta H_n T^{-1} \tag{13.87}$$

Correlations

Many equilibrium constants are measured in an attempt to answer such questions as: What makes an acid strong, or a complex weak? How is an equilibrium constant affected by changes in temperature, pressure, solvent or ionic medium? What is the effect on a complex of a minor change in the ligand, or of the replacement of one metal ion by another? Why are chelate complexes often, but not always, particularly stable? Although our understanding of these matters has developed considerably, we are still unable to predict values of equilibrium constants unless we know the values for extremely similar systems.

Since this book is concerned with methods of handling equilibria, rather than with interpreting the results obtained, we shall not discuss correlations in any detail. The numerous reviews available illustrate the interplay of broad trends of behaviour and individual exceptions.

It might seem that discussions should be based on values of $\Delta H_n^\ominus$ and $\Delta S_n^\ominus$ rather than on β_n or $\Delta G_n^\ominus$, but as enthalpies and entropies of complex formation are available for rather few systems, correlations are normally based on values of stability constants.

Many of the general trends are unsurprising. Metal ions which in their general chemistry show a preference for F^- or O^{2-} ions, for nitrogen atoms, or for heavier donors behave similarly towards complexing ligands. Thus Al^{3+} forms stable complexes with F^-; and Hg^{2+} with I^- and CN^-. Complexes of a given ligand with metal ions of one of these types become more stable as the ionisation potential, electronegativity and ionic charge increase, and as the ionic radius decreases. Stabilities of complexes of transition metal ions may be enhanced by ligand field stabilisation according to the electronic configuration of the ion and the strength of the ligand field. The stability of complexes of a particular metal ion with a number of very similar ligands increases with the basicity of the ligand unless complex formation is sterically hindered, or is affected by gross changes in the nature of the bonding. If complex formation involves decrease in charge, it is enhanced by a decrease in the dielectric constant of the solvent (e.g. by the addition of dioxan or an alcohol to a primarily aqueous solution). And whether or not the components are oppositely charged, stepwise constants normally decrease with the number of ligands added.

The study of chelates provides some more or less predictable generalisations. For a given number of donor atoms in a similar environment around the same metal ion, the stability increases with the number of chelate rings. Five-membered chelate rings are (usually) particularly stable. Predictably, metal ions form particularly stable complexes with those multidentate ligands which match their own stereochemistry. Thus Zn^{2+} favour polyamines in which the N-donors are tetrahedrally disposed, while Ni^{2+} prefers their planar analogues.

More remarkable, perhaps, are the individual exceptions to these trends. For example the stepwise stability constants of iron(II)-dipyridyl complexes fall in the unusual sequence $K_1 > K_2 \ll K_3$, probably because of change in electronic configuration (from $t_{2g}^4 e_g^2$ to t_{2g}^6) on addition of the third ligand. Despite Coulombic predictions, Ba^{2+} forms a more stable EDTA complex than Mg^{2+}, presumably because fewer of the six potential donor atoms can be accommodated around the smaller ion. Perhaps more interesting still are those instances of individuality which are as yet unexplained. How can we account for the wide variety of polynuclear hydroxo complexes? Why is K_2 greater than K_1 for silver ammonia complexes? Since a decrease in dielectric constant decreases the basicity of N-donor ligands, why does it have so little effect on the stability of metal–nitrogen complexes? Similar problems doubtless await future correlators of equilibrium constants as yet unmeasured, to be determined, perhaps, for systems under high pressures, or for equilibria in ionic melts.

Suggestions for further reading

No attempt has been made to give a comprehensive bibliography. The books and articles listed below have been selected with a view to the possible needs both of undergraduate students and of apprentice research workers. Between them, they provide a rich deposit of references to primary sources.

Methods for studying equilibria in solution

1. **F. J. C. Rossotti** and **H. S. Rossotti**, *The Determination of Stability Constants*, McGraw-Hill, New York (1961). A tough, generalised treatment of acids and metal complexes.
2. **M. T. Beck**, *The Chemistry of Complex Equilibria*, Van Nostrand Reinhold, London (1970). Deals with a wide range of metal complexes, including polynuclear, acido-, and other ternary, species.
3. **E. J. King**, *Acid–Base Equilibria*, Pergamon, London (1965). A full account, which includes the protonation of polyelectrolytes.
4. **J. E. Prue**, *Ionic Equilibria*, Pergamon, Oxford (1966). A readable treatment of metal complex formation.
5. **G. H. Nancollas**, *Interactions in Electrolyte Solutions*, Elsevier, Amsterdam (1966). Particularly useful for 'ion-pair' complexes.
6. **F. J. C. Rossotti**, in J. Lewis and R. G. Wilkins (eds), *Modern Coordination Chemistry*, Interscience, New York (1960). A predominantly interpretative review, containing a section on experimental techniques.
7. **H. S. Rossotti**, 'Design and publication of work on stability constants', *Talanta*, **21**, 809 (1974). A brief, critical summary.
8. **A. Albert** and **E. P. Serjeant**, *The Determination of Ionisation Constants*, Chapman and Hall, London (1971). A laboratory manual.

Graphical presentation of solution equilibria; and analytical applications

9. **L. G. Sillén**, in I. M. Kolthoff and P. J. Elving (eds), *Treatise on Analytical Chemistry*, Part I, Vol. I, Interscience, New York (1959).
10. **J. N. Butler**, *Ionic Equilibria*, Addison-Wesley, Reading, Mass. (1964). Contains an excellent account of log–log plots.

11. **W. B. Guenther**, *Chemical Equilibrium*, Plenum Press, New York (1975).
12. **L. Šůcha and St. Kotrlý**, *Solution Equilibria in Analytical Chemistry*, Van Nostrand Reinhold, London (1972).
13. **J. Inczédy**, *Analytical Applications of Complex Equilibria*, Ellis Horwood (Wiley), Chichester (1976).
14. **H. Freiser** and **Q. Fernando**, *Ionic Equilibria in Analytical Chemistry*, Wiley, London (1963).

Equilibria involving macromolecules

Ref. 3.
15. **C. Tanford**, *Physical Chemistry of Macromolecules*, Wiley, New York (1961).
16. **G. Weber**, in B. Pullman and M. Weissbluth (eds), *Molecular Biophysics*, Academic Press, New York (1965).
17. **J. Steinhardt** and **J. A. Reynolds**, *Multiple Equilibria in Proteins*, Academic Press, New York (1969).
18. **F. R. N. Gurd**, in S. J. Leach (ed.), 'Physical principles and techniques', in *Protein Chemistry*, Academic Press, New York (1970).
19. **R. Österberg**, 'The interaction of bovine serum albumin with zinc(II) ions', *Acta Chem. Scand.*, **25**, 3829 (1971).
20. **D. J. R. Laurence**, in B. Carroll (ed.), *Physical Methods in Macromolecular Chemistry*, Vol. 2, Marcel Dekker, New York (1972).
21. **G. Weber**, 'The binding of small molecules to proteins', *Adv. Protein Chem.*, **29**, 2 (1975).
22. **J. A. Marinsky**, 'Ion binding in charged polymers', *Co-ord. Chem. Rev.*, **19**, 125 (1976).

Polynuclear and ternary complexes

Refs. 1 and 2.
23. **R. Österberg,** 'Metal complexes of peptides and related compounds', *Acta Chem. Scand.*, **14**, 471 (1960); **19**, 1445 (1965). Cf. B. Sackar and T. P. A. Kruck, *Canad. J. Chem.*, **51**, 3541, 3555, 3563 (1973), and W. A. E. McBryde, ibid., p. 3573.
24. **L. G. Sillén**, 'Some recent results in hydrolytic equilibria', *Pure Appl. Chem.*, **17**, 55 (1968). (Polynuclear complexes.)
25. **M. Wozniak** and **G. Nowogrocki**, 'Méthode potentiométrique générale d'étude des complexes mixtes', *Bull. Soc. Chim. France*, 1974, 435 (and good refs. therein).
26. **C. F. Baes** and **R. E. Mesmer**, *The Hydrolysis of Cations*, Wiley, New York (1976). (Polynuclear complexes.)

Experimental techniques

Refs. 1, 2 and 3.
27. **H. Rossotti**, *Chemical Applications of Potentiometry*, Van Nostrand, London (1969).
28. **R. A. Durst** (ed.), *Ion Selective Electrodes*, National Bureau of Standards, Washington DC (1969).
29. **H. P. Bennetto** and **A. R. Willmott**, 'Electrochemical measurements with amalgam electrodes', *Quart. Rev. Chem. Soc.*, **25**, 501 (1971).
30. **R. G. Bates**, *Determination of pH*, Wiley, New York (1973).
31. **R. P. Buck**, 'Ion selective electrodes, potentiometry, and potentiometric titrations', *Anal. Chem.*, **46**, R28 (1974).
32. **L. Johannson**, 'Solubility measurements and complex formation in solution', *Co-ord. Chem. Rev.*, **3**, 293 (1968).
33. **D. H. Liem**, 'Application of LETAGROP to data for liquid–liquid distribution equilibria', *Acta Chem. Scand.*, **25**, 1521 (1971). Cf. ib., id., **22**, 753, 733 (1968).
34. **W. A. E. McBryde**, 'Spectrophotometric determination of equilibrium constants in solution', *Talanta*, **21**, 979 (1974).
35. **P. Gans** and **H. M. N. H. Irving**, 'The calculation of stability constants of weak complexes from spectrophotometric data', *J. Inorg. Nucl. Chem.*, **34**, 1885 (1972).
36. **G. D. Watt, J. J. Christenson** and **R. M. Izatt**, 'Thermodynamics of metal cyanides', *Inorg. Chem.*, **4**, 220 (1965).

Computational techniques

Ref. 1 for graphical methods; refs 24–26 and 33–35 on the use of computers.
37. **D. Dyrssen, D. Jagner** and **F. Wengelin**, *Computer Calculation of Ionic Equilibria and Titration Procedures*, Almquist and Wiksell, Stockholm (Wiley, London) (1968).
38. **C. W. Childs, P. S. Hallam** and **D. D. Perrin**, 'Applications of digital computers in analytical chemistry, Part II, Equilibrium constants', *Talanta*, **16**, 1119 (1969).
39. **F. J. C. Rossotti, H. Rossotti** and **R. J. Whewell**, 'The use of electronic computing techniques in the calculation of stability constants', *J. Inorg. Nucl. Chem.*, **33**, 2051 (1971).
40. **P. Gans**, 'Numerical methods for data-fitting problems', *Co-ord. Chem. Rev.*, **19**, 99 (1976).
41. **A. B. Calder** and **D. A. Calder**, *Nomograms for Chemists*, Royal Inst. Chem. London. Lecture Series 1962, No. 1. (See also ref. 10 and P. D. Lark, B. R. Craven and R. C. L. Bosworth, *The Handling of Chemical Data*, Pergamon, Oxford (1968).)

Compilations of data

42. **D. D. Perrin**, *Dissociation Constants of . . . in Aqueous Solution*, IUPAC: *Inorganic Acids and Bases*, 1969; also in *J. Pure Appl. Chem.*, **20**, 133 (1969).
Organic Acids, 1961; also in *J. Pure Appl. Chem.*, **1**, 190 (1961).
Organic Bases, 1965, and Supplement, 1972.
43. **L. G. Sillén** and **A. E. Martell**, *Stability Constants of Metal Ion Complexes*, Chem. Soc., London (1964; Supplement, 1970).

Index

Reference electrodes, 84, 96
  see also Calomel electrode
Replacement reactions, 135, 167, 181
Resistance, 94–5, 98
  see also Electrical Conductivity
Reversibility, of cells, 99
Ribonuclease, 58–9, 63

Salt bridge, 96–7
Sedimentation, 134
Selectivity, of glass electrode, 91
Self-association, 124
Shape parameter, 160
Sillén, L. G., 132, 150
Silver ammines, 5–7
Solubility, 33–4, 42, 47, 102–6, 113,
  178–81
  auxiliary ligand for, 105
  calculations involving, 178–81
  measurement of, 105
Solubility product, 102–5, 178–81
Solutions
  for potentiometry, 99–100
  preparation of, 77
  properties of, 115–23
Solvent extraction, *see* Liquid–liquid
  partition
Solvents, *see* Mixed solvents; Non-
  aqueous solvents
Spectrophotometry
  of dibasic acids, 47
  of metal complexes, 115–21
  of monobasic acids, 29–32, 35
  of polybasic acids, 54
  processing of measurements, 145–6,
    148, 150
  see also Beer–Lambert law
Stability constants
  calculation of, 141–53
  of metal complexes, 69–71
  of mixed complexes, 125
  of monobasic acids, 17
  of polynuclear complexes, 125
  overall, 37, 52, 69–71
  stepwise
    of dibasic acids, 37
    of polybasic acids, 52
    of metal complexes, 69–71
  trends in, 184–5
  uses of, 152–5

Standard state, 10–11
  in ionic medium, 36
  in water, 18
Stoichiometric coefficients, 12, 85, 88
Stoichiometric constants, *see*
  Concentration Quotients
Strong acids, *see* Acids, strong
Successive constants, *see* Stability
  constants stepwise
Successive extrapolation *see*
  Extrapolation methods

Tautomeric constant, 49
Temperature
  control of, 80
  effect on constants, 183–4*
Ternary species *see* Mixed complexes
Titrations
  for monobasic acids, 22–3
  for polynuclear complexes, 130
  for proteins, 137
  potentiometric, 85, 96, 99–100
Transfer, free energy of, 71
Transport number, 88–9, 93, 96
Tribasic acids, 52–7
TTA, 110
Two-phase systems, *see* Distribution,
  Heterogeneous

Ultraviolet spectrophotometry *see*
  Spectrophotometry
Uranium(VI), 69

Vanadium(IV), 69
Vapour pressure, 102, 121
Volhard titration, 5, 7

Washburn number, 88
Water
  as ligand, 68–71, 77
  dissociation of, 18, 172
  standard state of, 18, 70
Weak complexes, 71, 93
Weighting factors, 149

Zwitterion, 48